ESSENTIALS OF FOOD SANITATION

Norman G. Marriott

Virginia Polytechnic Institute and State University

Consulting Editor

Gill Robertson, M.S., R.D.

CHAPMAN & HALL

I T P® International Thomson Publishing

New York • Albany • Bonn • Boston • Cincinnati • Detroit • London • Madrid • Melbourne
Mexico City • Pacific Grove • Paris • San Francisco • Singapore • Tokyo • Toronto • Washington

JOIN US ON THE INTERNET

WWW: http://www.thomson.com

EMAIL: findit@kiosk.thomson.com

thomson.com is the on-line portal for the products, services and resources available from International Thomson Publishing (ITP). This Internet kiosk gives users immediate access to more than 34 ITP publishers and over 20,000 products. Through *thomson.com* Internet users can search catalogs, examine subject-specific resource centers and subscribe to electronic discussion lists. You can purchase ITP products from your local bookseller, or directly through *thomson.com*.

Visit Chapman & Hall's Internet Resource Center for information on our new publications, links to useful sites on the World Wide Web and an opportunity to join our e-mail mailing list. Point your browser to: **http://www.thomson.com** or **http://www.thomson.com/chaphall/foodsci.html** for Food Science.

Cover design: Andrea Meyer, emDASH inc.

A service of I**T**P®

Copyright © 1997 Chapman & Hall

Printed in the United States of America

For more information, contact:

Chapman & Hall
115 Fifth Avenue
New York, NY 10003

Chapman & Hall
2-6 Boundary Row
London SE1 8HN
England

International Thomson Publishing
Hirakawacho-cho Kyowa
Building, 3F
1-2-1 Hirakawacho-cho
Chiyoda-ku, 102 Tokyo
Japan

Thomas Nelson Australia
102 Dodds Street
South Melbourne, 3205
Victoria, Australia

Chapman & Hall GmbH
Postfach 100 263
D-69442 Weinheim
Germany

International Thomson Editores
Campos Eliseos 385, Piso 7
Col. Polanco
11560 Mexico D.F.
Mexico

International Thomson
Publishing Asia
221 Henderson Road #05-10
Henderson Building
Singapore 0315

1 2 3 4 5 6 7 8 9 10 XXX 01 00 99 98 97

Library of Congress Cataloging-in-Publication Data

Marriott, Norman G.
    Essentials of food sanitation / Norman G. Marriott; consulting
editor, Gill Robertson.
        p.   cm.—(Food science texts series)
    Includes bibliographical references and index.
    ISBN 0-412-08011-7 (alk. paper)
    1. Food industry and trade—Sanitation.   I. Robertson, Gill.
II. Title.   III. Series.
TP373.6.M36   1997   97-4088
664—dc21   CIP

British Library Cataloguing in Publication Data available

To order this or any other Chapman & Hall book, please contact International Thomson Publishing, 7625 Empire Drive, Florence, KY 41042. Phone: (606) 525-6600 or 1-800-842-3636. Fax: (606) 525-7778, e-mail: order@chaphall.com.

For a complete listing of Chapman & Hall titles, send your requests to Chapman & Hall, Dept. BC, 115 Fifth Avenue, New York, NY 10003.

*To my family, who have supported
my professional endeavors and provided
an abundance of love*

CONTENTS

PREFACE

This book is intended for anyone learning about or working in the food industry. It provides essential food sanitation information for everyone involved in food production, food processing, food preparation, and foodservice.

The first chapter of the book explains what sanitation means, why it is so important, and how it is regulated. The next three chapters provide background information about basic microbiology, how food becomes contaminated, and the importance of personal hygiene and good work habits. Five chapters cover various aspects of cleaning, sanitation, waste disposal, and pest control. A chapter highlights programs in quality assurance and hazard analysis critical control points. The next several chapters take a step-by-step look at sanitation in different food-processing, food production, and foodservice operations. The final chapter looks at how management affects sanitation.

The book includes a comprehensive glossary, and appendices provide a list of resource organizations and agencies, a summary of pathogenic organisms, and charts showing minimal internal cooking temperatures for meats, and maximum storage times for fresh and frozen foods.

Each chapter includes learning objectives, study questions, and ideas as to how students and workers can find out more about the topic. Throughout the book, boxes highlight real-life stories and data to help readers apply the information in the text.

Poor sanitation is expensive. Victims of foodborne disease have the expenses of medical bills and lost work hours. Food manufacturers or foodservice operators have immediate expenses of food recalls, wasted food, and investigations. But the company also suffers the long-term effects of bad publicity and loss of customers.

This book will help students and workers understand why food sanitation is so important in all aspects of the food industry. This understanding will help motivate them to use good sanitation on the job and help ensure the safety of our food supply.

ACKNOWLEDGMENTS

Appreciation is expressed to Gill Robertson for the written contributions that she provided to make this book a more pleasant reading experience.

ESSENTIALS
OF
FOOD
SANITATION

Sanitation: Definitions and Regulations

ABOUT THIS CHAPTER

In this chapter you will learn:

1. Why modern food production makes it even more important than before to use sanitary methods
2. Why a well-designed plant is not enough to ensure that food is safe
3. What sanitation means
4. How sanitation is regulated by government agencies
5. Why a planned sanitation program is essential

INTRODUCTION

Over the last decade, the food industry has grown tremendously. Sanitary practices have also changed and are now more complex. Food processing and preparation depend on more mechanized and large-volume processes. The foods produced by these processors and retailers are eaten by millions of people each day. Therefore, it is more and more important for workers to understand sanitary food-handling principles and food hygiene. Workers who understand why food sanitation is so important are more likely to use safe practices.

IMPORTANCE OF SANITATION

Plant Design and Sanitary Practices

Most food-processing plants are designed to be hygienic. But if proper sanitary methods are not used, food can still be contaminated by microorganisms. These microorganisms may cause the food to spoil, or the people who eat the food may become ill.

However, if hygienic methods are used, clean and safe foods can be produced, even in older plants with less-than-ideal designs. Food hygiene practices are at least as important as the design of the physical plant in producing safe food.

During this century, improved methods of food processing, preparation, and packaging have helped to improve the quality and safety of the food supply and to reduce the cost of food. Convenience foods and other processed foods still have some problems with food contamination and waste disposal, especially when large volumes of food are produced.

Training and Resources

Few colleges and universities provide training in food sanitation. Very few colleges offer even one course on this topic. Also, the resources available to foodservice and food production managers for training staff have been very limited. As part of a national campaign to reduce the risk of foodborne illness and to increase knowledge of food-related risks, the Food Safety and Inspection Service (FSIS) of the United States Department of Agriculture (USDA) and the Food and Drug Administration (FDA) of the United States Department of Health and Human Services (HHS) have established the USDA/FDA Foodborne Illness Education Information Center. The center provides information about prevention of foodborne illness to educators, trainers, and organizations developing education and training materials for food workers as well as consumers. The center maintains a database listing computer software, audiovisuals, posters, games, teaching guides, and training materials. See the end of this chapter for more information about this service.

Benefits of Food Sanitation Programs

A sanitation program is a planned way of practicing sanitation. Most owners or managers of food operations want a clean operation. They may not understand the principles of sanitation or the benefits of effective sanitation. But operators begin to understand the importance of food sanitation when they see how it can affect the profits of their operation. The benefits of a good sanitation program include the following:

1. *Complying with regulations.* Inspections are becoming stricter because inspectors are placing more emphasis on microbial and chemical hazards. Therefore, a good sanitation program is essential. Inspectors and regulatory agencies are more likely to trust operators who enforce sanitation programs.
2. *Preventing a catastrophe.* Thousands of cases of foodborne illness are reported each year. Because relatively few cases are reported, millions of cases actually occur. Because so many people eat food produced at one food operation, a single food sanitation problem may cause an outbreak involving many cases of foodborne illness.
3. *Improving quality and shelf life of foods.* Even if foods do not cause illness, poor food sanitation can cause food spoilage and bad color, smell, or flavor. Spoiled foods reduce sales and communicate to consumers that food safety is not a pri-

ority. A major national supermarket chain found that poor sanitation in meat operations caused increased labor, trim loss, and packaging costs leading to 5 to 10% lower profits.

4. *Reducing energy, maintenance, and insurance costs.* Dirty, clogged coils collect microorganisms, and blowers and fans spread them around the operation. Coil cleaning and sanitizing lower the risk of airborne contamination of foods. Clean coils exchange heat more efficiently and can reduce energy costs as much as 20%. Insurance carriers may give lower rates to a clean establishment because they are less likely to have claims for falls and accidents caused by greasy floors.

5. *Increasing quality and confidence.* A sanitation program can improve customer relations, reduce the risk to public health, and increase employee morale.

DEFINITIONS AND APPLICATION TO FOOD INDUSTRY AND FOODSERVICE

Definition of Sanitation

The word *sanitation* comes from the Latin word *sanitas,* which means "health." In the food industry, sanitation means creating and maintaining hygienic and healthful conditions. Scientific principles are used by healthy food handlers in a hygienic environment to produce wholesome food. Sanitation can reduce the growth of microorganisms on equipment and dirt on food. This can reduce contamination of food by microorganisms that cause foodborne illness and food spoilage.

Sanitation is more than just cleanliness. Food or equipment can be free of visible dirt and still be contaminated with microorganisms or chemicals that can cause illness or food spoilage. Sanitary principles also apply to waste disposal (see Chap. 8) and can help reduce pollution and improve ecological balance.

Sanitation is an applied science. Sanitation relates physical, chemical, biological, and microbial principles to food, the environment, and health. Some microorganisms cause food spoilage and foodborne illness, but others are beneficial in food processing and preparation. Sanitation scientists need to be able to control microorganisms so that they are beneficial rather than harmful.

The nutrients in food that humans need are also nutrients for the microorganisms that cause foodborne illness. Foods that are packaged for self-service in stores, especially fresh meats, have a large surface area exposed. Bacteria can grow rapidly in these foods. Therefore, effective sanitation programs must be used to keep these fresh foods safe and attractive for a reasonable period of time.

Examples of poor sanitation programs. Proper sanitation is important for food safety. The following examples show how poor hygiene can lead to outbreaks of foodborne illnesses:

• *Chunky.* Several years ago, an outbreak of foodborne illness was traced to Chunky chocolate candy with fruit and nuts. The manufacturer had to recall the product and remove candy bars from stores. It took a long time to solve the problem and restore consumer confidence in the product.

- *Starlac.* This was a leading brand of powdered milk until an outbreak of food-borne illness caused by *Salmonella* was traced to the product several years ago. The product was taken off the shelves, and the brand name was retired because of bad publicity.

- *Bon Vivant.* This company produced a line of canned gourmet foods until an outbreak of foodborne illness caused by *Clostridium botulinum* was traced to one of its products. The bad publicity forced the company to go out of business.

- *A supermarket.* A woman purchased chicken legs from the heated delicatessen case in a supermarket. When she ate the chicken 2 hours later, it tasted so bad that she ate only a mouthful. She vomited soon afterwards and had such severe symptoms that she had to be hospitalized. *Staphylococcus aureus* microorganisms were found in the chicken. The source of these microorganisms was never found, but the fact that just one mouthful could produce such severe illness shows the importance of hygienic practices.

- *U.S. Navy ship.* Shrimp salad caused an outbreak of foodborne illness on board a U.S. Navy ship, affecting 28 sailors. The food preparation facilities and methods were sanitary, but the employee who prepared the salad had a draining sore on his thumb. Drainage from the sore contained the same strain of *Staphylococcus aureus* microorganisms that had infected the salad.

- *Country club buffet.* A buffet served to approximately 855 people at a New Mexico country club was followed by an outbreak of acute gastrointestinal illness. *Staphylococcus aureus* phage type 95 was found in the turkey and dressing and in some of the food handlers' noses and stools. The food handlers were given training by state environmental personnel to prevent similar problems in the future.

Consumers have the right to expect wholesome and safe food products. Poor sanitation can lead to loss of sales and profits, damaged food products, loss of consumer confidence, bad publicity, and legal action.

CURRENT REGULATIONS

National vs. Local Laws

The United States currently has many regulations to control the preparation and manufacture of food. This chapter will not cover all the details of these rules. Instead, the major agencies involved with food safety and their primary responsibilities will be highlighted. Local regulations vary, so it is important to contact the local health department to find out what is required in your area.

Laws and Regulations

The requirements of the law are spelled out in regulations, which may change when problems or potential problems are found. Laws are passed by legislators and must be signed by the chief executive. After the law is passed, the agency that will enforce it prepares detailed regulations that spell out what the law or act will require. Regulations for food provide standards for building design, equipment design, levels of food additives, sanitary practices, food labels, and training for positions that require certification.

Development of regulations. Several steps have to be taken when regulations are developed from laws. For federal laws, the agency publishes proposed regulations in the *Federal Register.* Comments, suggestions, or recommendations can be sent to the agency within 60 days after the proposal is published. The comments are reviewed, and then final regulations are published. Dates by which the regulations must be carried out are included.

Types of regulations. There are two types of regulations: substantive and advisory. Substantive regulations are more important because they have the power of law and must be carried out. Advisory regulations are guidelines. Sanitation regulations are substantive because food must be made safe for the public. The use of *shall* in the wording of the regulations means a requirement; the word *should* means a recommendation.

Food and Drug Administration regulations. The FDA is part of the U.S. Department of Health and Human Services and is responsible for enforcing the Food, Drug, and Cosmetic Act as well as other laws. The FDA has had a dramatic influence on the food industry, especially in controlling adulterated foods. Any food that has been prepared or packed in an unsanitary way, that contains dirt, or that could be harmful to health for other reasons is adulterated. The act gives the FDA inspector the authority to enter and inspect any establishment where food is processed, packaged, or held; vehicles used to transport or hold food; and equipment, finished products, containers, and labels.

Adulterated or wrongly labeled products may be seized, or legal action may be taken against the food operation through an injunction or restraining order. The order is effective until the FDA is sure that the violations have been corrected. The FDA approves cleaning compounds and sanitizers for food plants by their chemical names, but not by their trade names. The following table gives examples of chemicals that are approved for cleaning and sanitizing:

Type of Sanitizer	Chemical Name
Bleach-type sanitizer	Sodium hypochlorite
Organic chlorine sanitizer	Sodium or potassium salts of isocyanuric acid
Quaternary ammonium products	*n*-Alkyldimethylbenzyl-ammonium chloride
Acid anionic sanitizer component	Sodium dodecylbenzenesulfonate
Iodophor sanitizers	Oxypolyethoxyethanol-iodine complex

The FDA regulations state the maximum concentration of each compound that can be used if they will not be rinsed off.

USDA regulations. The U.S. Department of Agriculture (USDA) is responsible for three areas of food processing, based on the Federal Meat Inspection Act, the Poultry Products Inspection Act, and the Egg Products Inspection Act. The agency that takes care of inspection is the Food Safety and Inspection Service (FSIS), which was established in 1981. The three acts give the USDA authority over operations in processing plants. The USDA has required that an inspector be present during all slaughtering and processing of red-meat animals and poultry. Some processors have experimented with a voluntary quality control program of self-inspection and record keeping. The facilities, equipment, sanitation practices, and programs must be approved for self-inspection.

Federal authority usually involves only foods that are shipped between states. But the three laws on meat, poultry, and eggs have extended USDA authority to within states if state inspection programs cannot provide proper controls.

The USDA authorizes cleaning and sanitizing compounds that can be used in federally inspected meat-, poultry-, rabbit-, and egg-processing operations. Categories are used to define how different cleaners, laundry compounds, sanitizers, hand-washing compounds, and water treatment compounds can be used.

Environmental regulations. The Environment Protection Agency (EPA) enforces many acts important to the food industry, including the Federal Water Pollution Control Act; Clean Air Act; Federal Insecticide, Fungicide, and Rodenticide Act (FIFRA); and the Resource Conservation and Recovery Act. The EPA also registers sanitizers by both their trade and chemical names.

VOLUNTARY SANITATION PROGRAMS

Voluntary programs give more responsibility to the food industry so that regulatory agencies can concentrate on other processes and operations. These programs also encourage constructive interaction between regulators and food producers and processors.

USDA Program

As the meat and poultry industry has become bigger and more complex, it has become more expensive for the USDA to inspect facilities continuously. In 1979, the USDA proposed a voluntary Total Quality Control (TQC) program designed to replace continuous inspection and constant supervision. This program has now been implemented in several processing plants. Plants must submit a signed agreement to participate, details of their quality control procedures, and characteristics of their organization. The USDA reviews records kept by the company, instead of maintaining continuous inspection of the facilities. The goal is to maintain cost-efficient inspections without reducing the safety and acceptability of the products.

FDA Programs

The FDA has a similar program for food-processing plants known as the Industry Quality Assurance Assistance Program (IQAAP). Under IQAAP, participating firms do not have to report any failures to meet approved quality assurance plans. This has helped to improve the relationship between the FDA and the food industry. The firm has to develop a Segment Quality Assurance Plan, which is reviewed by the FDA. Again, the FDA can audit records kept by the company.

Hazard Analysis Critical Control Points (HACCP)

This concept helps prevent food spoilage and contamination. It is endorsed by several organizations. At the moment, most regulary agencies consider HACCP to be voluntary, but will probably require it in all processing plants in the future. In fact, HACCP will be required in seafood, meat, and poultry plants in the future. This program is discussed in detail in Chapter 10.

ESTABLISHING SANITARY PRACTICE

The Sanitarian's Role

The employer is responsible for establishing and maintaining sanitary practices to protect public health and maintain a positive image. The sanitarian or food technologist in charge of sanitation oversees this process. He or she considers the impact of economic and aesthetic factors on sanitation. The sanitarian guards public health and counsels management on quality control and sanitary methods.

The Sanitation Department

A large food-processing company should have a separate sanitation department. This department should be given the same importance as production or research departments, and the chief sanitation administrator should be directly responsible to top management. Maintenance of sanitation should be separate from maintenance of production and machinery so that the sanitation department can monitor sanitary practices across the company. Production efficiency, quality control, and sanitary practice may appear to be three separate goals. But these three factors affect each other and should be coordinated.

Ideally, an organization should have a full-time sanitarian with assistants, but this is not always possible. Instead, a quality control technician, production foreman, or superintendent may be trained to manage sanitation. But unless the sanitarian has an assistant to take care of some of the day-to-day tasks, she or he may not have enough time to consider all the details, and the sanitation program may not be successful. Sanitation staff should have access to outside resources, such as a university, trade association, or private consultant, to avoid being influenced by the conflicting interests of different departments.

The Sanitation Program

A planned sanitation maintenance program is essential to meet legal requirements, to protect the reputation of the brand name and products, and to ensure that products are safe, of high quality, and free from contamination. All phases of food production and all areas in the plant should be included in the sanitation program, including equipment and floors. The program should begin by inspecting and monitoring raw materials that enter the facility, because these are potential sources of contamination.

The program should be comprehensive and critical. As problems are found, the ideal solution should be found, regardless of the cost. The whole operation can then be evaluated to find practical and economical ways to ensure safety and hygiene.

SUMMARY

* Sanitation means creating and maintaining hygienic and healthful conditions. It applies the sciences of microbiology, biology, physics, and chemistry to hygienic methods and conditions.
* Sanitary practices and hygienic conditions are more and more important because food is being processed and prepared in larger volumes than before.
* Food can be contaminated with microorganisms that cause spoilage or illness if sanitary methods are not used, even in well-designed plants.
* The Food, Drug, and Cosmetic Act covers all food except meat and poultry products from harvest through processing and distribution. Meat and poultry products are monitored by USDA. Pollution of air, water, and other resources is controlled by the EPA.
* Voluntary programs help firms take responsibility for sanitation without continuous inspection.
* A planned sanitation program is essential to meet regulatory requirements, protect the reputation of the brand name, and ensure that products are safe, of high quality, and free from contamination.

BIBLIOGRAPHY

Anon. 1976. *Plant Sanitation for the Meat Packing Industry.* Office of Continuing Education, University of Guelph and Meat Packers Council of Canada.

Clarke, D. 1991. FSIS studies detection of food safety hazards. *FSIS Food Safety Review* (Summer): 4.

Collins, W. F. 1979. Why Food Safety? In *Sanitation Notebook for the Seafood Industry,* G. J. Flick, Jr., et al., eds. Polytechnic Institute and State University, Blacksburg.

Environmental Protection Agency. 1972. Federal Environmental Pesticide Control Act of 1972 (amendment to FIFRA). U.S. Government Printing Office, Washington, D.C.

————. 1976. Resource Conservation and Recovery Act. U.S. Government Printing Office, Washington, D.C.

————. 1977. *Federal Water Pollution Control Act as Amended*. U.S. Government Printing Office, Washington, D.C.

————. 1978. *Amendments to Federal Insecticide, Fungicide and Rodenticide Act*. U.S. Government Printing Office, Washington, D.C.

Food and Drug Administration. *Federal Food, Drug and Cosmetic Act as Amended* (revised as amendments are added). U.S. Governmental Printing Office, Washington, D.C.

Guthrie, R. K. 1988. *Food Sanitation,* 3d ed., Chapman & Hall, New York.

Katsuyama, A. M. 1980. *Principles of Food Processing Sanitation.* The Food Processors Institute, Washington, D.C.

Marriott, N. G. 1994. *Principles of Food Sanitation,* 3d ed. Chapman & Hall, New York.

National Restaurant Association Educational Foundation. 1992. *Applied Foodservice Sanitation.* 4th ed. John Wiley & Sons, in cooperation with the Education Foundation of the National Restaurant Association, Chicago.

STUDY QUESTIONS

1. What does *sanitation* mean?
2. Are food sanitation regulations substantive or advisory? What does this mean?
3. Which agency is responsible for the safe production of most foods? Which foods are not monitored by this agency?
4. What are two benefits of voluntary inspection programs?
5. Give two reasons why firms need to have a separate department responsible for sanitation.
6. List three reasons why a planned sanitation program is essential.

TO FIND OUT MORE ABOUT SANITATION DEFINITIONS AND REGULATIONS

1. Call your local public health department (look in the Blue Pages of the telephone book), and ask them if they have information on local food sanitation regulations for food processors and foodservice operators.
2. Call the USDA Meat and Poultry Hotline, (800) 535-4555, 10:00 a.m. to 4:00 p.m. Eastern Time. Professional home economists will answer your questions about proper handling of meat and poultry, how to tell if they are safe to eat, and how to better understand meat and poultry labels. Ask for a list of their pamphlets and brochures.
3. Visit your library and look through recent issues of *FDA Consumer* magazine to find articles on food sanitation.
4. Call a local food production company (look in the Yellow Pages of the telephone book). Ask the sanitarian about its food sanitation program.
5. Contact the Foodborne Illness Education Information Center, (301) 504-5719 or e-mail: croberts@nalusda.gov, for information about the Foodborne Illness Educational Materials Database. A copy of the database is available on diskette in ASCII format. The database can also be accessed via the Internet.

GOOD SANITATION SAVES MONEY!

Foodborne illness is unpleasant, and food spoilage is wasteful. They are also very expensive for individuals and for organizations. Look at some of the costs that can be saved by good sanitation.

COSTS TO THE INDIVIDUAL

Medical bills
Lost work time
Wasted food

COSTS TO THE FOOD COMPANY

Wasted food
Bad publicity and loss of consumer confidence leading to lower sales
Lost production time
Legal fees
Higher insurance premiums

COSTS TO THE INDIVIDUAL'S EMPLOYER

Loss of productivity
Hiring replacement staff

CHAPTER 2

Microorganisms

INTRODUCTION

To understand food sanitation, you need to understand microorganisms that cause food spoilage and foodborne illness. Microorganisms are everywhere. Food spoilage happens when microorganisms spoil the color, flavor, or texture of food, and food-borne illness happens when people eat harmful microorganisms or their toxins along with the food. Good sanitation can prevent food spoilage and food-poisoning microorganisms from growing and causing damage.

COMMON MICROORGANISMS

What Are Microorganisms?

A microorganism is a microscopic form of life found on anything that has not been sterilized. Microorganisms use nutrients, get rid of waste products, and reproduce.

About 100 years ago, Louis Pasteur and his coworkers realized that microorganisms cause disease and found that heat could destroy them.

Microorganisms and Food

Most foods spoil easily because they have the nutrients that microbes need to grow. Food is acceptable for a longer time, and foodborne illness is less likely, if the growth of microbes is controlled. Proper sanitation during food processing, preparation, and serving controls food spoilage and pathogenic (disease-causing) bacteria.

The most common microorganisms in food are bacteria and fungi. Fungi (molds or yeasts) are less common than bacteria. Molds are multicellular (organisms with several cells), and yeasts are usually unicellular (single-cell organisms). Bacteria usually crowd out fungi and are unicellular. Viruses are usually carried from person to person rather than via food, but they can be transferred among unhealthy employees. Parasites are not common in processed or cooked foods. While they can be foodborne, they do not multiply in food.

Molds

Molds have a variety of colors and look fuzzy or cottonlike. Molds can develop tiny spores that ride in the air and are spread by air currents. If these land on a new food and the conditions allow the spores to germinate, they grow. Molds can survive wider ranges of temperature and pH than bacteria and yeasts . Molds prefer pH 7.0 (neutral), but can tolerate pH 2.0 (acid) to 8.0 (alkaline). Molds prefer room temperature to refrigerator temperatures, but can even grow below freezing temperature. Most molds prefer a minimum water activity (A_w) of about .90, but some molds can grow at A_w as low as .60. (See later in this chapter for more about water activity.) At an A_w of .90 or higher, bacteria and yeasts usually grow better and crowd out molds. When the A_w is below .90, molds are more likely to grow than are bacteria and yeasts. Pastries, cheese, and nuts are low in water content and are more likely to spoil from mold than from bacteria and yeasts.

Yeasts

Yeasts have larger cells than bacteria and divide by producing buds. They multiply more slowly than bacteria: each yeast generation takes about 2 to 3 hours. It takes about 40 to 60 hours for food to be spoiled if it starts with one yeast organism per gram of food. Like molds, yeasts can be spread by air and also by hands, equipment, and surfaces. Yeast colonies usually look moist or slimy and are a creamy white color. Yeasts prefer an A_w of .90 to .94, but can grow below .90. A few can even grow at an A_w as low as .60. Yeasts like a somewhat acid pH, between 4.0 and 4.5. Yeasts are most likely to grow on vacuum-packaged foods with an acid pH. Growth of yeasts often makes foods smell fruity and feel slippery.

Bacteria

Bacteria are about 1 μm in diameter (0.00004 in). Their shape may be short rods (bacilli), round (cocci), or oval. Different types of bacteria group together in different ways. Some round bacteria form clusters like a bunch of grapes, e.g., staphylococci. Some rods or spheres link up to form chains, e.g., streptococci. Some round bacteria join together in pairs (diploid formation), e.g., pneumococci. Some form a group of four (tetrad formation), e.g. sarcina. Bacteria make various pigments that may be yellow, red, pink, orange, blue, green, purple, brown, or black. Bacteria may discolor foods, especially those that have unstable color pigments, such as meat. Some bacteria discolor food by forming slime. Some bacteria produce spores that can survive heat, chemicals, and other harsh treatments. Some of these bacteria produce toxins that can cause foodborne illness.

Viruses

Viruses are infectious and are one-tenth to one-hundredth the size of bacteria. Most viruses can be seen only by using an electron microscope. Viruses can only grow inside the cells of another organism and live inside bacteria, fungi, plants, and animals. One virus that has caused several outbreaks of disease in restaurants in the past 10 years is hepatitis. Infectious hepatitis A can be carried by food that has been poorly handled. The symptoms are nausea, cramps, and, sometimes, jaundice, and can last from several weeks to several months. Raw shellfish from polluted waters are a common source of hepatitis. Foods that are handled a lot and those that are not heated after they are handled—such as sandwiches, salads, desserts, and ice—are most likely to transmit viruses. Hepatitis A is very contagious, so food handlers must wash their hands thoroughly after using the toilet, before handling food and eating utensils, after touching animals, and after diapering, nursing, or feeding infants.

GROWTH PHASES OF MICROORGANISMS

Microbial cells have a growth cycle with five phases: lag phase, logarithmic growth phases, stationary growth phase, accelerated death phase, and reduced death phase. Figure 2.1 shows a typical microbial growth curve.

Lag Phase

The lag phase is an adjustment or adaptation period. After the food has been contaminated, microbes take a while to get used to the environment. At first, the stress of the new location may cause a small decrease in the number of microbes. But then they begin to grow slowly. Low temperature extends the lag phase. Cold temperatures reduce growth of microorganisms so that each generation takes longer to multiply. Decreasing the number of microbes that contaminate food, equipment, or buildings also slows their growth. Good sanitation and hygiene lower the initial

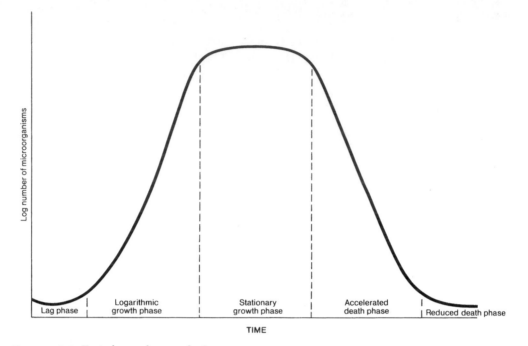

FIGURE 2.1. Typical growth curve for bacteria.

number of microorganisms and can also help to extend the lag phase. Figure 2.2 shows how temperature and contamination levels can affect growth of microbes.

Logarithmic Growth Phase

Bacteria multiply by binary fission, which means that all the cell parts duplicate inside the cell, and then the cell splits into two identical daughter cells. The logarithmic growth phase may last for 2 hours to several hours. The number of microorganisms and environmental factors, such as nutrients and temperature, affect the logarithmic growth rate. Because the population doubles with each generation, and a generation may be short as 20 minutes, it is possible for one cell to become 262,144 cells in just 6 hours!

Stationary Growth Phase

Growth slows down when the environment becomes less ideal because of competition for nutrients, temperature changes, buildup of waste products, or other microbes growing in the same food. This phase usually lasts for between 24 hours and 30 days, depending on the nutrient supply and other factors in the environment.

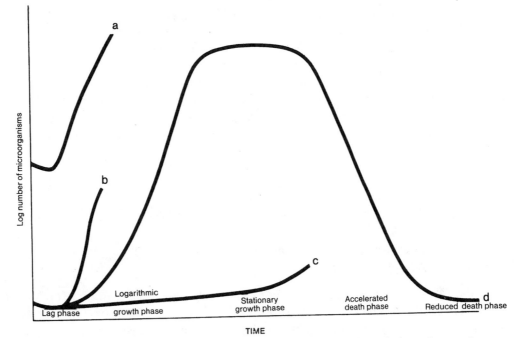

FIGURE 2.2. The effect of amount of initial contamination and length of lag phase on the growth curve of microorganisms: (a) high initial contamination and poor temperature control (short lag phase), (b) low initial contamination and poor temperature control (short lag phase), (c) low initial contamination and rigid temperature control (long lag phase), and (d) typical growth curve.

Accelerated Death Phase

Lack of nutrients, buildup of waste products, and competition from other microbe populations eventually cause rapid death of microbial cells. This phase also lasts between 24 hours and 30 days, depending on the temperature, nutrient supply, type of microbe, age of the microorganisms, use of sanitizers, and other microorganisms in the food.

Reduced Death Phase

When the accelerated death phase reduces the population number, the death rate slows down. After this phase, the organism is decomposed or sterilized, or another microorganism breaks it down.

WHAT CAUSES MICROBES TO GROW?

Temperature

Different microbes each have optimum, minimum, and maximum growth temperatures. Each type of microorganism grows fastest at its optimum temperature, but still grows at any temperature between its minimum and maximum. The further the temperature is from the optimum temperature for that microorganism, the more slowly it grows. The environmental temperature controls the growth rate and which microorganisms will grow best. For example, a change in temperature of only a few degrees may encourage different organisms to grow in a food. Therefore, small temperature changes can cause different types of food spoilage and foodborne illness. Controlling the temperature is an important way to reduce microbial growth.

Most microorganisms grow best between 14°C (58°F) and 40°C (104°F), although some will grow below 0°C (32°F), and others will grow at temperatures up to 100°C (212°F). Sometimes microbes are classified by their optimal growth temperature:

1. *Thermophiles* prefer hot temperatures above 45°C (113°F), e.g., *Bacillus stearothermophilus, Bacillus coagulans,* and *Lactobacillus thermophilus.*
2. *Mesophiles* prefer medium temperatures between 20°C and 45°C (68°F and 113°F), e.g., most lactobacilli and staphylococci.
3. *Psychrotrophs* tolerate cold temperatures below 20°C (68°F), e.g., *Pseudomonas* and *Moraxella-Acinetobacter.*

Each of these three categories includes some bacteria, molds, and yeasts. Molds and yeasts tend to prefer cooler temperatures than bacteria. When the temperature is below about 5°C (40°F) (refrigerator temperature), microorganisms grow slowly, and many pathogens (disease-causing microbes) stop growing.

Oxygen

Some microorganisms must have oxygen to grow, some grow only without oxygen, and some grow either with or without oxygen. Microorganisms that require oxygen are called *aerobic* microorganisms (e.g., *Pseudomonas* spp.). Microbes that grow without oxygen are called *anaerobic* microorganisms (e.g., *Clostridium* spp.). Those that grow with or without oxygen are called *facultative* microorganisms (e.g., *Lactobacillus* spp.).

Relative Humidity

All microorganisms need water to grow. If the air is humid, water vapor condenses on food, equipment, walls, and ceilings. Moist surfaces encourage microbes to grow. If the air is dry (low relative humidity), microbes are less likely to grow. Bacteria need a higher relative humidity than yeasts and molds. The best relative humidity level for bacteria is 92% or more. Yeasts prefer 90% relative humidity or more. Molds thrive best when the relative humidity is between 85% and 90%.

Water Activity (A_w)

Microbes need water. If less water is available, microbial growth is slower. However, the total amount of water does not affect microbial growth—it is the amount of water that is actually available for the microbes to use that is important. Water activity (A_w)

is the amount of water that is available to microorganisms. Many microorganisms prefer an A_w of .99, and most need an A_w higher than .91 to grow. Relative humidity (RH) and A_w are related (RH = $A_w \times 100$). Therefore, an A_w of .95 is about the same as an RH of 95%. Water activity refers to the availability of water in a food or beverage. Relative humidity refers to the availability of water in the atmosphere around the food or beverage. Most food products have an A_w of about .99. Bacteria need a high A_w, molds need a lower A_w, and most yeasts are between the two. Most spoilage bacteria cannot grow if A_w is below .91, but molds and yeasts can grow when A_w is .80 or lower. Therefore, molds and yeasts are more likely to grow on dried-out foods or surfaces than are bacteria.

In some climates, a high relative humidity may hydrate (dampen) the top layer of foods and allow microorganisms to grow, even if the rest of the food has a low water activity. Good packaging stops the humid air from reaching the food so that microorganisms cannot start to grow.

pH

The pH value shows how acid or alkaline something is; pH ranges from 0.0 to 14.0 (the number is log_{10} of the reciprocal of the hydrogen ion concentration). pH 7.0 is neutral, pH values below 7.0 are acid, and pH values above 7.0 are alkaline. Most microorganisms prefer a neutral pH. Yeasts can grow in acid and do best when pH is 4.0 to 4.5. Molds can tolerate a wider range of pH (2.0–8.0), although they usually grow best in an acid pH. Molds can grow when conditions are too acidic for bacteria and yeasts. Bacteria grow best when the pH is close to neutral. But some acid-loving (acidophilic) bacteria can grow on food or debris at a pH as low as 5.2. Below pH 5.2, microbial growth is very slow.

Nutrients

Microbes need more than water and oxygen. They also need sources of energy (carbohydrates, fats, or proteins), nitrogen, minerals, and vitamins to grow. Most microbes get nitrogen from amino acids (broken-down proteins) or other chemicals that contain nitrogen (e.g., urea). Molds can use whole proteins, carbohydrates, and fats because they have enzymes that can digest them. Many bacteria can also break down nutrients, but most yeasts need nutrients in simple, broken-down forms. All microorganisms need minerals, but different microbes need different amounts and kinds of vitamins. Molds and some bacteria can make enough B vitamins for themselves, but other microorganisms need a supply of B vitamins.

Inhibitors

Inhibitors slow the growth of microbes. Substances or things that slow the growth of bacteria are called *bacteriostats*. Those that kill bacteria are called *bactericides*. Sometimes food processors add bacteriostatic substances, such as nitrites, to foods to discourage growth of bacteria. Most bactericides are used to decontaminate foods or to sanitize equipment, utensils, and rooms after they have been cleaned (see Chap. 6).

Biofilms

Biofilms were discovered in the mid-1970s. They are very small colonies of bacteria that stick to a surface using a matrix or glue that traps nutrients and other microorganisms. A biofilm is like a beachhead for bacteria and is not easy to get rid of with sanitizers. When a microorganism lands on a surface, it uses filaments or tendrils to attach itself. The organisms produce a sticky substance that cements the bacteria in place within a few hours. A biofilm builds up several layers of the matrix, populated with microbes such as *Salmonella, Listeria,* and *Pseudomonas.* Eventually the biofilm becomes a tough layer that has to be scraped off. A surface can be cleaned and sanitized, but a well-established biofilm has layers of microorganisms that are protected from the sanitizer. It is easy for chunks of the outer layers of the biofilm to come off when food rubs against it. This contaminates the food.

Biofilms are very hard to remove during cleaning. *Pseudomonas* and *Listeria monocytogenes* are microbes that cause problems because they are protected by biofilms. Heat seems to be more effective than chemical sanitizers for destroying biofilms. Also, it is easier to clear biofilms off Teflon than off stainless-steel surfaces. Water-soluble sanitizers—such as caustics, bleaches, iodophors, phenols, and quaternary ammonium sanitizers—do not get into biofilms. Government regulations and industry guidelines do not cover removal of biofilms. Cleaning employees may need to use a sanitizer at 10 to 100 times the normal strength for it to work. It is very important to rinse the equipment or area carefully after using these high concentrations of sanitizers so that sanitizer residues do not contaminate the food.

Relationship Between Microbial Load, Temperature, and Time

At lower temperatures, the generation interval (time for one bacterial cell to become two cells) is longer. At temperatures below 4°C (40°F), the generation interval is very long. Figure 2.2 shows how temperature affects microbial growth. For example, freshly ground beef usually contains about 1 million bacteria per gram (a gram is about 1/28 of an ounce). When the population reaches about 300 million per gram, the meat becomes slimy and smells "bad." If the meat is more contaminated to begin with or if the temperature is high enough so that bacteria multiply rapidly, the food will spoil more quickly. For example, ground beef that contains 1 million bacteria per gram can be stored for about 28 hours at 15.5°C (60°F) before it is spoiled. At normal refrigerator temperatures (−1–4°C, 30–40°F), it can be stored for more than 96 hours.

HOW FOOD IS SPOILED

Food is spoiled when it is unfit for humans to eat. Spoilage usually means that microorganisms have decomposed or putrefied the food. Food that is unsafe does not necessarily look or smell spoiled.

Physical Changes

Physical changes are usually more obvious than chemical changes. Spoilage causes physical characteristics such as color, texture, thickness, smell, and flavor to change. Food spoilage is either aerobic or anaerobic, depending on whether oxygen is available. Aerobic spoilage by molds is normally only on the surface of the food, where oxygen is available. Moldy surfaces of foods such as meats and cheeses can be trimmed off, and the rest can be safe to eat. If bacteria grow extensively on the surface, toxins may get inside the food. Anaerobic spoilage happens in sealed containers or in the middle of foods where oxygen is not available. Facultative or anaerobic bacteria spoil food by souring, putrefaction, or tainting. When foods go sour, the microorganisms may release gases that make cans or containers swell.

Chemical Changes

The foods themselves (and the microorganisms that contaminate them) contain enzymes that break down proteins, fats, carbohydrates, and other large molecules into smaller and simpler chemicals. As the number of microbes increases, they start to break down the food. The texture of the food may become soft or slimy and the color may change. Without oxygen, protein may be broken down to a variety of sulfur-containing compounds, which usually smell offensive (like the smell of a bad egg).

FOODBORNE ILLNESS

The United States has the safest food supply in the world. But microorganisms in food still cause about 25 million cases of foodborne illness and 16,000 deaths each year. In the United States, the annual cost of foodborne illness and death averages about $3,000 per person, or about $75 billion. Each death related to foodborne illness costs about $42,300, including insurance and other expenses.

Foodborne Disease

A foodborne disease is any illness caused by eating food. A foodborne disease outbreak is when two or more people have the same illness, which usually involves the stomach and/or intestines, after eating the same food. Bacteria cause about two-thirds of all foodborne illness outbreaks. The incidence of foodborne diseases is not known because so many cases are not reported. But between 6 and 50 million Americans probably become ill each year because of microorganisms in food. *Salmonella* spp. alone causes between 200,000 and 1 million cases each year. About 9,000 people die each year from foodborne illness, and the cause is not known in about 60% of the 200 foodborne disease outbreaks that are reported each year.

Salmonella spp., *Staphylococcus aureus, Clostridium perfringens, Clostridium botulinum, Listeria monocytogenes, Yersinia enterocolitica, Escherichia coli, Vibrio* spp., and *Campylobacter* spp. are all carried in food and could be responsible for illness. Many different home-cooked and commercially prepared foods have caused outbreaks, but foods of animal origin—such as poultry, eggs, red meat, seafood, and dairy products—are most often responsible.

Food Poisoning

It is helpful to understand the names of different types of foodborne illness.

- *Food poisoning* is an illness caused by eating food that contains toxins made by microbes or chemical poisons.
- *Food intoxication* is food poisoning caused by toxins from bacteria.
- *Chemical poisoning* is food poisoning caused by chemicals (such as polychlorinated biphenyls) that are in the food by accident. (Illnesses caused by microorganisms are more common than those caused by chemicals.)
- *Food infections* are illnesses that are not caused by toxins produced by bacteria but are due to eating infectious microorganisms, such as bacteria, viruses, or parasites.
- *Food toxicoinfections* are foodborne illnesses caused from a combination of food intoxication and food infection. Disease-causing bacteria grow in the food, and the patient eats large numbers with the food. The pathogen continues to grow in the intestine and produces a toxin that causes illness symptoms.
- *Psychosomatic food illness* is illness caused by the mind after seeing someone else sick or after seeing a foreign object, such as an insect or rodent, in a food product.

Food producers need up-to-date knowledge of production, harvesting, and storage techniques to evaluate the quality and safety of raw materials. They must understand the design, construction, and operation of food equipment to be able to control the processing, preservation, preparation, and packaging of food products. Food products are vulnerable to contamination and need to be safeguarded to prevent food poisoning.

Staphylococcal Food Poisoning

This is caused by eating the toxin produced by *Staphylococcus aureus*. This facultative, non-spore-forming microorganism produces an enterotoxin that causes irritation of the stomach and intestines. Staphylococcal food poisoning is rarely fatal, but it can cause vomiting and diarrhea. If it does cause death, it is usually because the patient also has other illnesses. The microorganisms that cause staphylococcal food poisoning are very common and can be carried by healthy people. About half of humans are carriers. Food is usually contaminated when infected food handlers touch it. Potato salad, custard-filled pastries, dairy products (including cream), poultry, cooked ham, and tongue are the most common food sources of staphylococcal food poisoning. If the temperature is ideal and the food is heavily contaminated, staphylococci can multiply enough to cause food poisoning before the color, flavor, or aroma of the food changes. Heating at 66°C (150°F) for 12 minutes destroys *Staphylococcus aureus* organisms, but the toxin they produce is not destroyed by normal cooking or processing.

Salmonella Food Poisoning

Salmonellosis (foodborne illness from *Salmonella*) usually causes nausea, vomiting, and diarrhea, because the toxins irritate the walls of the intestines. About 1 million of these microorganisms have to be eaten by healthy people to cause an infection. It takes fewer organisms to cause salmonellosis in people who are already ill. The time between eating the food and developing symptoms is generally longer than for staphylococcal food poisoning. Salmonellosis rarely causes death, but deaths may occur if the patient is an infant, elderly, or already sick from other illnesses. Patients with acquired immune deficiency syndrome (AIDS) are susceptible to this foodborne illness.

Although salmonellae can be found in animal products, most meat is contaminated by handlers during processing. Heat processes that kill *Staphylococcus aureus* will destroy most species of *Salmonella*. Salmonellosis can usually be blamed on poor sanitation and temperature abuse, because most strains of salmonellae cannot grow at refrigerator temperatures.

Clostridium perfringens Food Poisoning

Clostridium perfringens multiplies rapidly in meat items that have been cooked, allowed to cool slowly, and then held for a long time before serving. Large numbers of active bacteria have to be eaten for this foodborne illness to occur.

Proper sanitation and refrigeration of foods at all times, especially of leftovers, can usually prevent an outbreak of *C. perfringens* foodborne illness. Living microorganisms are destroyed when leftover foods are reheated to 60°C (140°F).

Botulism

Botulism is a true food poisoning caused by eating a toxin produced by *Clostridium botulinum* when it grows in food. The toxin is extremely potent (the second most powerful biological poison for humans) and affects the victim's peripheral nervous system. About 60% of those infected die from respiratory failure. Those who recover may need respiratory treatment for the rest of their lives. Table 2.1 shows the symptoms, incubation time, foods involved, and prevention of botulism and other common food poisonings.

Because *C. botulinum* can be found in soil, it is also found in water. Therefore, seafoods are more common sources of botulism than are animal foods. But the most common sources of botulism are home-canned vegetables and fruits with a low-to-

TABLE 2.1. Characteristics of Foodborne Illnesses

Illness	Causative Agent	Symptoms	Average Time Before Onset of Symptoms	Foods Usually Involved	Preventive Measures
Botulism	Toxins produced by *Clostridium botulinum*	Impaired swallowing, speaking, respiration, and coordination; dizziness and double vision	12–48 hours	Canned low-acid foods, including canned meat and seafood, smoked and processed fish	Proper canning, smoking, and processing procedures; cooking to destroy toxins, proper refrigeration and sanitation
Staphylococcal foodborne illness	Enterotoxin produced by *Staphylococcus aureus*	Nausea, vomiting, abdominal cramps due to gastroenteritis (inflammation of the lining of the stomach and intestines)	3–6 hours	Custard- and cream-filled pastries, potato salad, dairy, ham, tongue, and poultry	Pasteurization of susceptible foods, proper refrigeration and sanitation
Clostridium perfringens foodborne illness	Toxin produced by *Clostridium perfringens*	Nausea, occasional vomiting, diarrhea, and abdominal pain	8–12 hours	Cooked meat, poultry, and fish held at nonrefrigerated temperatures for long periods of time	Prompt refrigeration of unconsumed cooked meat, poultry, or fish; maintain proper refrigeration and sanitation
Salmonellosis (food infection)	Infection produced by ingestion of any of over 1,200 species of *Salmonella* that can grow in the gastrointestinal tract of the consumer	Nausea, vomiting, diarrhea, fever, abdominal pain; may be preceded by chills and headache	6–24 hours	Insufficiently cooked or warmed-over meat, poultry, eggs, and dairy products; these are especially susceptible when kept refrigerated for a long time	Cleanliness and sanitation of handlers and equipment, pasteurization, proper refrigeration and packaging

Table 2.1. (Continued)

Illness	Causative Agent	Symptoms	Average Time Before Onset of Symptoms	Foods Usually Involved	Preventive Measures
Ttrichinosis (infection)	Trichinella spiralis (nematode worm found in pork)	Nausea, vomiting, diarrhea, profuse sweating, fever, muscle soreness	2–28 days	Insufficiently cooked pork and products containing pork	Thorough cooking of pork to an internal temperature of 59°C (138°F) or 77°C (150°F) or higher with microwave cooking: freezing storage of uncooked pork at minus 15°C (60°F) or lower, for a minimum of 20 days; avoid feeding pigs raw garbage
Aeromonas foodborne illness	Aeromonas hydrophila	Gastroenteritis	—	Water, poultry, red meats	Sanitary handling, processing, preparation, and storage of foods; store foods below 2°C (35°F)
Campylobacteriosis foodborne illness	Campylobacter spp.	Diarrhea, abdominal pain, cramping, fever prostration, bloody stools, headache, muscle pain, dizziness, and, rarely, death	3–5 hours	Poultry and red meats	Sanitary handling, processing, preparation, and storage of muscle foods
Listeriosis	Listeria monocytogenes	Meningitis or meningo-encephalitis, listerial septicemia (blood poisoning), fever, intense headache, nausea,	1–7 days	Milk, coleslaw, cheese, ice cream, poultry, red meats	Avoid consumption of raw foods and contact with infected animals; store foods below 2°C (35°F)

Table 2.1. (Continued)

Illness	Causative Agent	Symptoms	Average Time Before Onset of Symptoms	Foods Usually Involved	Preventive Measures
		vomiting, lesions after direct contact, collapse, shock, coma; mimics influenza, interrupted pregnancy, stillbirth; 30% fatality rate in infants and immuno-comprised children and adults			
Yersiniosis	*Yersinia enterocclitica*	Abdominal pain, fever, diarrhea, vomiting, skin rashes for 2–3 days, and, rarely, death	24–36 hours	Dairy products, raw meats, seafoods, fresh vegetables	Sanitary handling, processing, preparation, and storage of foods
Escherichia coli 0157:H7 infection	Enterohemmornagic *Escherichia coli* 0157:H7	Hemorrhagic colitis, hemolytic uremic syndrome with 5–10% acute mortality rate, abdominal pain, vomiting, anemia, thrombocytopenia, acute renal injury with bloody urine, seizures, pancreatitis	—	Ground beef, dairy products, raw beef, water, apple cider, mayonnaise	Sanitary handling, irradiation, cooking to 65°C (149°F) or higher

medium acid content. Recent outbreaks have involved improperly held baked potatoes and grilled onions and some flavored oils. Because *C. botulinum* is anaerobic, it can also contaminate canned and vacuum-packaged foods that are sealed off from oxygen. It is important not to eat foods in swollen cans because the gas produced by the organism causes the swelling. However, some strains of *C. botulinum* do not produce gas. Heating to at least 83°C (180°F) for 30 minutes during processing is enough to protect smoked fish.

Proper sanitation, proper refrigeration, and thorough cooking are essential to prevent botulism. The toxin is destroyed by heat, but the bacterial spores are only destroyed by severe heat treatment. Heating at 85°C (185°F) for 15 minutes destroys the toxin. Table 2.2 shows the combinations of temperatures and times needed to destroy the spores completely.

TABLE 2.2. Temperatures and Times Required to Destroy Completely *Clostridium botulinum* Spores

Temperature, °C	°F	Time, min
100	212	360
105	220	120
110	230	36
115	240	12
120	248	4

Campylobacter

The intestines of wild and domestic animals often contain this microbe. *Campylobacter* is now the greatest cause of foodborne illness in the United States. It is one of the most frequent causes of bacterial diarrhea and other illnesses, and it may also cause ulcers. *Campylobacter* causes veterinary diseases in poultry, cattle, and sheep, and is quite common on raw poultry. *Campylobacter* lives in the intestines of cattle, sheep, swine, chickens, ducks, and turkeys. Animal feces contain the microorganism, so muscle foods (meat and poultry) can be contaminated during slaughtering if sanitary methods are not used. Milk, eggs, and water that have been in contact with animal feces may also contain *C. jejuni*. Red meat is less likely to contain *C. jejuni* than is poultry. It is not likely that *Campylobacter* spp. can be eliminated from domestic animals because it is so widespread.

Campylobacteriosis (foodborne illness from *Campylobacter*) is at least twice as common as salmonellosis. Each year the 4 million cases in the United States cost over $2 billion. The symptoms of foodborne illness from *Campylobacter* vary. People with a mild case may not have any signs of illness but excrete the microorganism in their feces. People with a severe case may have muscle pain, dizziness, headache, vomiting, cramping, abdominal pain, diarrhea, fever, prostration, and delirium. Patients may find blood in their stool after 1 to 3 days of diarrhea. The illness usually lasts from 2 to 7 days. Death is rare, but it can happen.

Campylobacter outbreaks are most common in children over 10 years old and in young adults, although all age groups are affected. Symptoms and signs of *C. jejuni* infection are similar to those caused by other illnesses of the stomach and intestines. Symptoms may begin between 1 and 11 days after eating contaminated food, but illness usually develops 3 to 5 days after eating this microbe.

Sanitary handling and cooking animal foods properly are the best ways to control *Campylobacter.* Heating contaminated foods to an internal temperature of 60°C (140°F) and holding them at this temperature for several minutes for beef and about 10 minutes for poultry easily destroys *C. jejuni.*

Listeria

The Centers for Disease Control have estimated that 2,000 cases of listeriosis (food-borne illness caused by *Listeria* spp.) occur in the United States each year. A survey during 1992 found that this illness causes about 425 deaths per year.

L. monocytogenes lives in the intestines of over 50 domestic and wild species of animals and birds, including sheep, cattle, chickens, and swine, as well as in soil and decaying vegetation. This microorganism is also found in the air, stream water, sewage, mud, trout, crustaceans (e.g., crabs), houseflies, ticks, and the intestines of human carriers who have symptoms. This pathogen has been a problem in most foods, from chocolate and garlic bread to dairy products, meat, and poultry. Processed milk, soft unpasteurized cheeses, and other dairy products are the most common food sources of *L. monocytogenes,* but vegetables can also be contaminated if they have been fertilized with manure from infected animals. Elimination of *Listeria* is not practical and may be impossible, but food producers and consumers can control its survival.

This microbe grows best at 37°C (98°F), but it can grow at temperatures between 0°C (32°F) and 45°C (113°F). Therefore, it can grow in the refrigerator. It grows twice as fast at 10°C (50°F) as at 4°C (40°F) and will survive freezing, but is usually destroyed at temperatures above 61.5°C (142°F).

Listeriosis is most common in pregnant women, infants, adults over 50 years old, people who are sick from another disease, and other people with reduced resistance or immunity. In adults, the most common symptoms are meningitis or meningoencephalitis. Symptoms may be mild and influenza-like. If a pregnant woman is infected with listeriosis, the fetus may also be infected. This may cause a miscarriage or a stillborn child. If the infant survives, it may be born with septicemia or may develop meningitis. About 30% of newborn infants who get listeriosis die, and about 50% die if they get the infection in the first 4 days after birth.

Listeriosis is very dangerous to persons with acquired immunodeficiency syndrome (AIDS). AIDS severely damages the immune system, so patients are more likely to get a foodborne illness such as listeriosis.

Listeria monocytogenes can stick to food contact surfaces and form a biofilm that is hard to remove during cleaning. These microbes can contaminate food ingredients used in food processing, so the pathogen is constantly brought into the processing plant. A Hazard Analysis Critical Control Point (HACCP) and other quality assurance practices are the best ways to control this pathogen.

This microorganism is usually passed on by eating contaminated food, but it can also be passed by person-to-person contact. People who have been in contact with infected animals, soil, or feces may develop sores on their hands and arms.

About two-thirds of home refrigerators may contain *Listeria* spp. The best way to prevent listeriosis is to avoid ingesting raw (unpasteurized) milk, raw meat, and foods made from contaminated ingredients. Pregnant women should be especially careful to avoid contact with infected animals. Food processors must use a strict sanitation program and HACCP program to control *Listeria*.

Yersinia enterocolitica

Yersinia enterocolitica lives in the intestines and feces of wild and domestic animals. It can also live in raw foods of animal origin and nonchlorinated water from wells, streams, lakes, and rivers. This microorganism also seems to be transmitted from person to person.

Y. enterocolitica can multiply in the refrigerator, but more slowly than at room temperature. This microorganism is destroyed at temperatures over 60°C (140°F). When it is found in processed foods, it is usually occurs because they have been contaminated after cooking. *Y. enterocolitica* has been found in raw or rare red meats and poultry; dairy products such as milk, ice cream, cream, eggnog, and cheese curds; most seafoods; and fresh vegetables.

Not all types of *Y. enterocolitica* cause illness. Yersiniosis (foodborne illness caused by *Yersinia enterocolitica*) is most common in children and teenagers, although adults can get it too. The symptoms (fever, abdominal pain, and diarrhea) normally begin 2 to 7 days after eating the contaminated food. Patients may also vomit and have skin rashes. The abdominal pain caused by yersiniosis is similar to appendicitis. In the past, some children have had their appendix removed because of an incorrect diagnosis.

The illness normally lasts 2 to 3 days, although patients may have mild diarrhea and abdominal pain for 1 to 2 weeks. Death is rare, but can happen if the patient has complications. Proper sanitation in food processing, handling, storage, and preparation is the best way to prevent yersiniosis.

Escherichia coli 0157:H7

Recently, this pathogen has caused serious outbreaks of hemorrhagic colitis and hemolytic uremic syndrome. This pathogen is found in dairy products, water, apple cider, mayonnaise, and raw beef, but the most common cause of outbreaks is undercooked ground beef. Dairy cows can carry *E. coli* 0157:H7. Feces of cattle may contain this microorganism and can contaminate meat during processing. It is important to monitor and control slaughtering procedures and meat-processing operations to prevent this pathogen from growing. Beef should be cooked to an internal temperature of 70°C (158°F) to destroy this pathogen. A rigid sanitation program is essential to reduce foodborne illness outbreaks from this microorganism.

Other Bacterial Infections

Other bacterial infections cause foodborne illness. The most common of these infections is caused by *Streptococcus faecalis*. Another common cause of infections is *Escherichia coli*. This can cause "traveler's diarrhea," which can be a problem for people from developed countries when they visit developing countries with poor hygiene. Evisceration (removing the intestines) and cold storage of chickens at 3°C may allow *Aeromonas hydrophila* to grow. Water used to chill the birds and the process of removing the intestines may spread bacteria during broiler processing, which may be why *A. hydrophila* is so common on retail cuts of poultry.

Bacillus cereus is another common bacteria that can cause foodborne illness. *B. cereus* foodborne illness causes diarrhea and abdominal pain beginning 8 to 12 hours after the contaminated food is eaten and lasting for about 12 hours. It may also cause vomiting within 1 to 5 hours after eating contaminated food. *B. cereus* produces the toxin in the food before it is eaten. Heat does not easily destroy the toxin. Rice or fried rice from restaurants or warmed-over mashed potatoes that have been cooled slowly before reheating have caused outbreaks. The best way to control this foodborne illness is to use proper sanitation. Because *B. cereus* spores are airborne, food should be covered during holding, when possible. Restaurants should hold cooked starchy foods above 50°C (122°F) or cool them to less than 4°C (40°F) within 2 hours of cooking to prevent the bacteria from growing and producing the toxin.

Mycotoxins

Mycotoxins are produced by many fungi. The diseases caused by mycotoxins are called *mycotoxicoses*. Mycotoxicoses are not common in humans, but one mycotoxin (aflatoxin) seems to be linked to liver cancer in susceptible populations. Large doses of aflatoxins are very toxic, causing liver damage, intestinal and internal bleeding, and death. Mycotoxins get into food when mold grows on the food or when food producers use contaminated ingredients in processed foods.

Molds that can produce mycotoxins often contaminate foods. *Aspergillus, Penicillium, Fusarium, Cladosporium, Alternaria, Trichothecium, Byssochlamys*, and *Sclerotinia* are important in the food industry. Most foods can be infected by these or other fungi at some point during production, processing, distribution, storage, or display. If mold grows, it may produce mycotoxins. However, not all moldy foods contain mycotoxins. Foods that are not moldy are not necessarily free of mycotoxins, because the toxin can still be there after the mold has disappeared.

Aflatoxin is the most hazardous mycotoxin for humans. It is produced by *Aspergillus flavus* and *A. parasiticus*. These molds are found almost everywhere, and their spores are easily carried by air currents. They are often found in cereal grains, almonds, pecans, walnuts, peanuts, cottonseed, and sorghum. The microorganisms do not grow unless the food is damaged by insects, not dried quickly, or stored somewhere damp.

Aflatoxicosis causes loss of appetite, lethargy, weight loss, neurological problems, convulsions, and sometimes death. It can also cause liver damage, fluid buildup (edema) in the body cavity, and bleeding in the kidneys and intestines.

The best way to eliminate mycotoxins from foods is to prevent mold growth, insect damage, and mechanical damage during production, harvesting, transporting, processing, storage, and marketing. Mycotoxins are not produced when A_w is below .83 (about 8 to 12% moisture in grain). Therefore, grains should be dried quickly and completely and stored somewhere dry. Photoelectric eyes can examine grain and automatically remove discolored kernels that may contain aflatoxins. This equipment is used in the peanut industry to avoid difficult, boring, and expensive hand sorting.

Are There More Pathogens in Food Than There Used to Be?

Pathogens are found in food more often than in the past, but it may be because scientists have developed better methods to detect pathogens, rather than because foods are more contaminated. Other reasons why pathogens are found more often include the following:

1. *Changes in eating habits.* "Organic" products may seem healthy but can be unsafe. An outbreak of listeriosis in Canada was linked to coleslaw that was made from cabbage fertilized with sheep manure.
2. *More awareness of hazards, risks, and hygiene.* Better collection of data and use of computers have helped public health officials recognize foodborne listeriosis.
3. *Changes in the population.* Ill people are kept alive much longer. These people are more likely to be infected. Tourists and immigrants may spread certain diseases worldwide.
4. *Changes in food production.* Raw materials are often produced in large amounts. Large batches can easily create places for microorganisms to grow (e.g., because the center takes longer to cool to refrigerator temperature after cooking). A large batch of food can spread contamination even further when it is used as ingredients in other foods.
5. *Changes in food processing.* Vacuum packages and cold storage allow facultative microorganisms and psychrotrophs to survive.
6. *Changes in food storage.* Food processors and consumers can store foods such as vegetables, salads, soft cheeses, and meats for longer in the refrigerator, but this can allow growth of psychrotrophic pathogens such as *Listeria monocytogenes*.
7. *Changes in microorganisms.* Over time, the genetic makeup of microorganisms can change. New strains may develop that have characteristics from different bacteria. For example, a pathogen may become able to grow in cold temperatures, survive cooking, or grow in more acid or alkaline foods.

DESTROYING MICROBES

Regardless of how microorganisms are destroyed, they die at a relatively constant rate in the accelerated death phase (Fig. 2.1). Some things can change the death rate, such as a lethal agent or a mixed population of sensitive and resistant cells. Heat, chemicals, and radiation can all destroy microbes.

Heat

Heat is the most common method of killing spoilage and pathogenic bacteria in foods. Heat processing is a way to cook food and destroy microorganisms at the same time. Researchers work hard to find the ideal temperature and length of cooking time to destroy harmful microorganisms without overcooking the food. The amount of time it takes to completely sterilize a liquid containing bacterial cells or spores at a given temperature is the *thermal death time* (TDT). The TDT depends on the microorganism, the number of cells, and what it is growing on.

Another way to measure destruction of microbes is *decimal reduction time* (D value). This is the number of minutes it takes to destroy 90% of the cells at a given temperature.

Chemicals

Many chemical compounds that destroy microorganisms should not be used to kill bacteria in or on food. Food processors use chemicals to sanitize equipment and utensils that could contaminate food. These chemicals are rinsed off, so they cannot contaminate food. Sanitizing using heat has become more expensive, so the food industry uses chemical sanitizers more often.

Radiation

Microbiologists do not fully understand how radiation destroys microbes. It seems to inactivate parts inside the cell as they absorb its energy. When radiation inactivates a cell, it cannot divide and grow.

INHIBITING GROWTH OF MICROBES

Food processors use mild versions of methods used to kill microorganisms to inhibit or slow down growth of microbes. Heating at lower temperatures or for shorter times, low-dose irradiation, or treatment with diluted chemicals (such as food preservatives) can injure microorganisms and slow their growth without killing them. Injured microorganisms have a longer lag phase; do not tolerate extreme heat, cold, acid, or alkali so well; and are more sensitive to less-than-ideal conditions. Microbes are more easily inhibited by two inhibitors at the same time, such as irradiation plus heat, or heat plus chemicals.

Freezing

Freezing and thawing kills some microbes. Those that survive freezing cannot grow during frozen storage. But freezing is not a practical way to reduce the number of microbes. Also, once the food is thawed, the microorganisms that survive freezing will grow as quickly as those that have not been frozen. Refrigeration works well with other methods of inhibition, such as preservatives, heat (pasteurization), and irradiation.

Chemicals

Chemicals that reduce A_w to a level that prevents growth of most bacteria can control bacteria. Salt and sugar are examples of chemicals that reduce A_w. When they are used to preserve foods, they are used in much greater amounts than those used to flavor and season foods. Chemical preservatives also prevent growth of microbes in foods by reducing the pH, altering the A_w, or interfering with the microbe's metabolism.

Dehydration

Dehydration also reduces the A_w to a level that prevents microbial growth. Dehydration is most effective when foods are also treated with another inhibitor, such as salt or refrigeration.

Fermentation

Fermentation produces desirable flavors and can control microbial growth. Fermentation produces acid that lowers the pH of the food. A pH below 5.0 restricts growth of spoilage microorganisms. Foods that are fermented and heated can be canned or bottled to prevent spoilage by aerobic yeasts and molds (e.g., pickles).

TESTS TO DETECT AND COUNT MICROBES

Various methods can detect growth of microbes in foods. Food scientists use different methods depending on what they need to know, the type of food, and the type of microbe. The American Public Health Association (APHA) and the Association of Official Analytical Chemists (AOAC) have developed standard testing procedures. It is very important to collect representative samples (see Chap. 10). Microbial tests are less accurate and precise than chemical tests. Therefore, food scientists need technical knowledge about microbiology and food products to choose the right test and interpret the results.

Although microbial analysis may not give exact results, it can show whether the sanitation program is working to keep food products safe and equipment, utensils, floors, walls, and other areas clean. Microbial tests can also predict shelf life or how long a food can be kept. Here is a brief summary of the most common microbial tests.

Total Plate Count

This counts the total population of aerobic microorganisms on equipment or food products. The total plate count method assesses contamination from the air, water, equipment surfaces, facilities, and food products. The microbiologist swabs the equipment, walls, or food products. He or she washes the material off the swab into a culture medium (such as standard-methods agar or plate count agar) that supports the growth of all microorganisms. The microbiologist dilutes the sample several times, depending on how many microbes are expected, and places it on a growth medium in a sterile covered plate (petri dish). The incubation temperature may be the same as the storage temperature of the food or the same temperature as the location of equipment or utensils (e.g., 4°C [40°F]) for refrigerator shelves, room temperature for a knife). The result is the number of colonies that grow. This does not show the type of microorganisms, although some colonies look different. Sometimes the bacteria are incubated on special growth media that allow only certain microorganisms to grow and be counted.

Press Plate Technique

This is also called the *contact plate technique.* It is similar to the plate count technique without the swabbing. The microbiologist opens a covered RODAC plate (similar to a petri dish) and presses the growth medium (agar) against the area to be sampled. The incubation process is the same as for the total plate count method. This method is easy and has less chance of error and contamination. The biggest problem is that it can only be used for flat surfaces that are lightly contaminated because the sample is not diluted. Sanitarians use press plates to check whether the sanitation program is effective. The number of colonies that grow show the amount of contamination.

Indicator and Dye Reduction Tests

Many microorganisms produce enzymes during their normal growth. These enzymes can react with indicators, such as dyes. The enzymes react with the dye and cause a color change. The speed of the reaction shows the number of microorganisms in the sample. Sometimes a dye-soaked filter paper is placed right onto a food sample or piece of equipment. The time it takes for the filter paper to change color shows the number of microbes. The biggest problem with this technique is that it does not show the exact amount of contamination. But it is quicker and easier than plate counts and can be a good tool to check the effectiveness of a sanitation program. There are many different test kits for specific organisms.

Direct Microscopic Count

The microbiologist dries and fixes a measured sample of food on a microscope slide. He or she stains it and counts the number of bacteria. Most stains do not distinguish between bacteria that are alive or dead, so this method shows the total number of microorganisms in a sample. This method gives some information about the type of bacteria in the sample, and the slides can be kept for future reference, but it is not used much because it is easy for analysts to make counting errors, and the method can examine only a small quantity.

Most Probable Number (MPN)

This common method estimates bacterial populations by lining up several tubes of liquid growth medium and adding a sample that has been diluted by different amounts to each tube. If bacteria grow, the medium looks cloudy. The number of microorganisms is shown by how much the sample had to be diluted to prevent growth. This method measures only live bacteria, and the microbiologist can do more tests on the bacteria that grow to identify them.

Impedance Measurements

Impedance can measure microbial metabolism, which is an indirect way to count the number of microbes in a sample. Impedance is the total resistance to the flow of an alternating electric current as it passes through something. Microbiologists measure microbial growth by looking at the changes in impedance measurements over 5 hours. In the future, sanitarians may use impedance as a rapid method of counting microbial load, although it can be used only when the food contains more than 100,000 microbes per gram. At the moment, the equipment costs about $70,000, and each sample costs about $2. The equipment can handle 128 to 512 samples at one time.

Direct Epifluorescence Filter Technique (DEFT)

This is a rapid, direct method of counting microorganisms in a sample. Cells are stained with a fluorescent dye that stains live and inactive bacteria different colors. English researchers developed this method to monitor milk samples, but food scientists also use it for other foods. Sanitarians use DEFT to check dairy foods and meat, beverages, water, and wastewater. Each test takes about 25 minutes, and each sample costs about $1, although costs vary and are based on the quantity purchased.

Salmonella 1-2 Test

This is a rapid screening test for *Salmonella*. If a dark band (immobilization band) forms in the medium, motile *Salmonella* have reacted with flagellar antibodies, and the test is positive.

This test uses a clear plastic device with two chambers. The chambers contain two different media. One contains flagellar antibodies to *Salmonella,* and the sample is added to the other. After about 4 hours of incubation, motile *Salmonellae* move from one chamber to the other and come in contact with the flagellar antibodies. The immobilization band forms after 8 to 14 hours.

CAMP Test

This is a test for *L. monocytogenes*. The sanitarian streaks the sample next to or across a streak of a known bacterium on a blood agar plate. Where the two streaks meet, the metabolic by-products of the two bacteria blend and react to break down the blood (hemolysis) in the plate medium.

Fraser Enrichment Broth/Modified Oxford Agar

Fraser broth encourages growth of *Listeria* and prevents growth of other microorganisms. A specially prepared U-shaped tube contains Fraser broth on both sides and modified Oxford agar in the middle. The sample is placed into one side of the tube. *Listeria* are the only microorganisms that grow and migrate through the modified Oxford agar into the other side of the tube. The pure culture causes black deposits in the Oxford agar and cloudiness in the second branch of Fraser broth.

SUMMARY

- Microbiology is the study of microscopic forms of life.
- Sanitarians need to understand how microorganisms cause food spoilage and foodborne illness so that they can handle foods in a sanitary way.
- Microorganisms cause food spoilage by changing how food looks, tastes, and smells. Foodborne illness happens when people eat food containing pathogenic microorganisms or their toxins.
- A sanitation program controls the number of microorganisms on equipment, processing plants, and food.
- Microorganisms have a growth pattern that looks like a bell curve. They grow and die at a logarithmic rate. The factors that affect growth of microbes include temperature, oxygen availability, water availability, pH, nutrients, and inhibitors.
- Microorganisms produce enzymes that break down proteins, fats, carbohydrates, and other molecules in food into simpler compounds. This breakdown causes food spoilage. Microorganisms that cause foodborne illness include *Staphylococcus aureus, Salmonella* spp., *Clostridium perfringens, Clostridium botulinum, Campylobacter* spp., *Listeria monocytogenes, Yersinia enterocolitica,* and molds that produce mycotoxins.
- The best ways to destroy microbes are heat, chemical sanitizers, and irradiation. The best ways to inhibit growth of microbes are refrigeration, dehydration, and fermentation.
- Sanitarians use several tests to measure the number and type of microbes in food samples and on equipment or other surfaces. These tests show whether the sanitation program is working.

BIBLIOGRAPHY

Felix, C. W. 1992. CDC sidesteps listeria hysteria from *JAMA* articles. *Food Protection Report.* 8(5):1.

Gillespie, R. W. 1981. Current status of foodborne disease problems. *Dairy, Food, and Environ. Sanit.* 1:508.

Gravani, R. B. 1987. Bacterial foodborne diseases. *Dairy, Food, and Environ. Sanit.* 7:137.

Longree, K., and Armbuster, G. 1996. *Quantity Food Sanitation,* 5th ed. John Wiley, New York.

Marriott, N. G. 1994. *Principles of Food Sanitation,* 3d ed. Chapman & Hall, New York.

Mascola, L., Lieb, L., Chiu, J., Fannin, S. L., and Linnan, M. J. 1988. Listeriosis: an uncommon opportunistic infection in patients with acquired immunodeficiency syndrome. *Am. J. Med.* 84:162.

National Restaurant Association Educational Foundation. 1992. *Applied Foodservice Sanitation,* 4th ed. John Wiley, New York. In cooperation with the Education Foundation of the National Restaurant Association, Chicago.

Niven, C. F., Jr. 1987. Microbiology and parasitology of meat. In *The Science of Meat and Meat Products,* p. 217. Food and Nutrition Press, Westport, Conn.

Seideman, S. C., Vanderzant, C., Smith, G. C., Hanna, M. O., and Carpenter, Z. L. 1976. Effect of degree of vacuum and length of storage on the microflora of vacuum packaged beef wholesale cuts. *J. Food Sci.* 41:738.

Zottola, E. A. 1972. *Introduction to Meat Microbiology.* American Meat Institute, Chicago.

STUDY QUESTIONS

1. Name the most common types of microorganisms in food, and give one example of each.
2. What do yeasts and molds look like on food?
3. Describe the five phases of microbial growth.
4. Which microorganisms grow best when water activity is low?
5. Name five factors that affect the growth of microorganisms.
6. Why are biofilms a problem in the food industry?
7. What is the difference between physical and chemical food spoilage?
8. Name four bacteria that commonly cause foodborne illness.
9. Which foods are likely to be contaminated with aflatoxin? How should foods be stored so that molds do not produce aflatoxin?
10. What is the difference between destroying and inhibiting microbes?

TO FIND OUT MORE ABOUT MICROORGANISMS IN FOOD

1. Call USDA's Meat & Poultry Hotline, (202) 472-4485, and ask for information about harmful microorganisms that grow in food.

2. Conduct an informal survey among your friends and family. Has anyone had a foodborne illness? Do they know what food caused the illness? What were the symptoms? How long did the illness last? Do they know what microorganism caused the illness? Was the illness reported to CDC?

3. Call a local fast-food chain, and ask what it does to protect its customers from foodborne illness such as *Escherichia coli* in undercooked ground beef.

4. Check your kitchen at home. Are all foods stored at the right temperature? Are foods covered to protect them from insects, moisture, and microorganisms in the air? Use the kitchen quiz "Can your kitchen pass the food safety test?" by Paula Kurtzweil in FDA Consumer, October 1995.

Sources of Food Contamination

ABOUT THIS CHAPTER

In this chapter you will learn:

1. Where microorganisms in foods come from
2. The difference between foodborne infection and foodborne intoxication
3. How microbes in food cause illness in people
4. How specific foods and ingredients become infected
5. How contamination of foods can be prevented and controlled

INTRODUCTION

Nutrients for Microorganisms

Food products provide nutrients that people need. Microorganisms need many of the same nutrients, so food products are an ideal source of nutrition for microorganisms. Foods also generally have a pH value (acidity level) that encourages growth of microorganisms.

Where Do Microorganisms in Food Come From?

Microorganisms are everywhere: in soil, feces, air, and water. Food can be contaminated any time it comes in contact with these substances during harvesting, processing, distribution, and preparation.

Animal carcasses. The intestines of animals used for meat are full of microorganisms, and even animals that seem healthy may have microorganisms in their liver, kidneys, lymph nodes, and spleen. It is easy for these microorganisms to reach the meat during slaughter and butchering. As the meat is cut into retail portions, more and more surface area is exposed to the microorganisms.

Controlling Growth of Microbes

Food processors try to control the growth of these microorganisms by controlling the temperature (refrigeration) and oxygen and moisture levels (wrapping and packaging) during processing and distribution of foods.

Temperature. Refrigeration is one of the most important ways contamination of foods is controlled. Refrigeration prevents outbreaks of foodborne illness by slowing the growth of microbes. But when foods are not handled properly in cold storage, they can become heavily contaminated. The growth rate of microorganisms can increase tremendously with only a slight increase in temperature.

Food volume and container size. Foods tend to cool slowly in air, and large containers or large volumes of food take a long time to cool through to the center. The surface of the food can feel well chilled while the middle is still warm. Slow cooling of large pieces of meat or large containers of broth in the refrigerator has caused many outbreaks of foodborne illness caused by growth of *Clostridium perfringens*.

Infection vs. Intoxication

There are two types of foodborne illness caused by microorganisms: infection and intoxication.

- Foodborne *infection* occurs when the microorganism is eaten with the food and multiplies in the person's body. Examples include *Salmonella, Shigella,* and some types of enteropathogenic *Escherichia coli*.
- Foodborne *intoxication* occurs when microorganisms grow and release toxins into the food before it is eaten. When the food is eaten, the toxins cause illness. Examples include *C. perfringens* and some strains of enteropathogenic *E. coli*.

THE CHAIN OF INFECTION

A small number of harmful microorganisms (pathogens) in food are unlikely to cause a foodborne illness. However, these few pathogens could cause a problem under certain conditions. One model that illustrates how various factors and events can be linked together to cause an infection is called the "chain of infection."

A chain of infection is made up of four links: agent, source, mode of transmission, and host. To cause a foodborne infection, each of these links must be present in the environment in which the food is produced, processed, or prepared.

- The *agent* is the pathogen that causes the illness.

- The *source* is where the pathogen comes from.
- The *mode of transmission* is how the pathogen is carried from the source to the food.
- The *host* is the food that supports growth of the pathogen.

To survive and grow, the pathogen also needs nutrients, moisture, the right level of acidity or alkalinity (pH) and oxidation-reduction potential, and lack of competitive microorganisms and inhibitors. The food also needs to be held in the best temperature range for that organism long enough to allow it to grow enough to cause infection or intoxication.

The chain of infection model shows that foodborne diseases have many causes. A disease agent (pathogen) has to be present, but each of the other steps is also essential in causing foodborne disease.

HOW FOODS BECOME CONTAMINATED

The food itself is the most common source of contamination. Equipment and waste products are other common sources.

Dairy Products

The udders of cows and milking equipment can contaminate milk products, although equipment with well-designed sanitary features and control of disease in dairy cows have made dairy products more wholesome. Pasteurization of milk products in processing plants has also reduced pathogens in milk. However, dairy products can be cross-contaminated by items that have not been pasteurized. Cross-contamination occurs when utensils or equipment used for unpasteurized milk are used for pasteurized milk, or when staff working with unpasteurized milk products move to an area containing pasteurized milk products. Not all dairy products are pasteurized, so some pathogens (especially *Listeria monocytogenes*) have become more common in the dairy industry. (See Chap. 11 for more information.)

Red-Meat Products

The muscles of healthy animals are nearly free of microorganisms while alive. Meat is contaminated by microorganisms on the animal's surfaces that have external contact (hair, skin, intestines, and lungs). While the animal is alive, its white blood cells and antibodies control infection. But these defense mechanisms are lost during slaughter.

Microorganisms first reach the meat if contaminated knives are used to bleed animals. The blood is still circulating and quickly carries these microorganisms throughout the animal's body. Microorganisms reach the surface of the meat when it is cut, processed, stored, and distributed. Meat can also be contaminated if it comes in contact with the hide, feet, manure, dirt, and visceral (intestinal) contents if the digestive organs are punctured. (See Chap. 12 for more information.)

Poultry Products

Poultry may be contaminated by *Salmonella* and *Campylobacter* during processing. These microorganisms are easily spread from one carcass to another during defeathering and removal of the intestines (evisceration). *Salmonellae* can also be transferred from contaminated hands, gloves, and processing tools. (See Chap. 12 for more information.)

Seafood Products

Seafoods may be contaminated with microbes during harvesting, processing, distribution, and marketing. Seafoods are excellent sources of proteins, B vitamins, and a number of minerals that bacteria need to grow. Therefore, microbes grow well on or in seafoods. Seafoods are handled a lot from the time they are harvested until they are eaten, which provides many opportunities for contamination. They may be also sometimes be stored without being refrigerated, which allows microorganisms to grow. (See Chap. 13 for more information.)

Ingredients

Ingredients (especially spices) can carry harmful or potentially harmful microorganisms and toxins. The amounts and types of these microbes and toxins depends on where and how the ingredient was harvested and how the ingredient was processed and handled. The food plant management team needs to know the type of hazards that can occur with each ingredient. Food processors should only obtain materials from suppliers that use good practices.

OTHER SOURCES OF CONTAMINATION

Equipment

Equipment can be contaminated during production and while it is not being used. Most equipment is designed to be hygienic, but it can still collect microorganisms and other debris from the air, employees, and food ingredients. Food is less likely to be contaminated if equipment is designed to be hygienic and is cleaned regularly and thoroughly.

Employees

The most common source of microorganisms in foods is employees. The hands, hair, nose, and mouth carry microorganisms that can be transferred to food during processing, packaging, preparation, and service by touching, breathing, coughing, or sneezing. Because the human body is warm, microorganisms grow and multiply rapidly. Therefore, sanitary practices, such as good handwashing and use of hairnets and disposable plastic gloves, are essential. (See Chap. 4 for more information.)

Air and Water

Water is used for cleaning and as an ingredient in many processed foods. However, if the water is not pure, it can contaminate foods. If the water source is contaminated, another source should be used, or the water should be treated by chemicals, ultraviolet units, or other methods.

Microorganisms in the air can contaminate foods during processing, packaging, storage, and preparation. The best ways to reduce air contamination are to use filters for air entering food-processing and preparation areas and to package or cover food products to reduce contact with air.

Sewage

Raw, untreated sewage can carry microorganisms, causing typhoid and paratyphoid fevers, dysentery, and infectious hepatitis. Raw sewage may contaminate food and equipment through faulty plumbing. If raw sewage drains or flows into drinking-water lines, wells, rivers, lakes, and ocean bays, the water and seafood will be contaminated. To prevent this kind of contamination, toilet facilities and septic tanks should be separated from wells, streams, and other water sources. Raw sewage should not be used to fertilize fields where fruits and vegetables are grown. (See Chap. 8 for more information.)

Insects and Rodents

Food and food waste attract flies and cockroaches to kitchens, foodservice operations, food-processing facilities, toilets, and garbage. These insects transfer dirt from contaminated areas to food through their waste products; mouth, feet, and other body parts; and saliva. Any pests should be eradicated and prevented from entering food-processing, preparation, and serving areas.

Rats and mice carry dirt and disease with their feet, fur, and feces. They transfer dirt from garbage dumps and sewers to food or food-processing and foodservice areas. (See Chap. 9 for more information.)

HOW TO PREVENT AND CONTROL CONTAMINATION OF FOODS

The Environment

Foods should not be touched by human hands if they will be eaten raw or after they have been cooked. If contact is necessary, workers should thoroughly wash their hands before handling the food and wash regularly during handling or use disposable plastic gloves. During storage, holding, and service, processed and prepared foods should be covered with a clean cover that fits well and will not collect loose dust, lint, or other debris. If the food cannot be covered, it should be held in an enclosed dust-free cabinet. Foods in single-service wrappers or containers, such as milk and juice, should be served directly from these containers. Foods on a buffet should be served on a steam table or ice tray and protected by a transparent shield from sneezes, coughs, and other contact with air, employees, and customers. Food that touches an unclean surface should be thoroughly washed or thrown away. Equipment and utensils used in food processing, packaging, preparation, and service should be cleaned and sanitized between use. Foodservice managers should train employees to handle dishes and eating utensils so that their hands do not touch any surface that will come into contact with food or the consumer's mouth.

During Storage

Storage facilities should have plenty of space and an organized storage layout, and stock should be rotated. These measures help to reduce contamination from dust, insects, rodents, and dirt, and allow for easy cleaning. Storage area floors should be swept or scrubbed, and shelves or racks should be cleaned and sanitized. (See Chaps. 5, 6, and 7 for more information.) Trash and garbage should not be allowed to accumulate in food storage areas.

Litter and Garbage

The food industry generates large amounts of waste from used packaging, containers, and waste products. Refuse should be kept in appropriate containers and removed from the food area regularly. The best method (required by some regulatory agencies) is to use separate containers for food waste from those used for litter and rubbish. Clean, disinfected, seamless trash containers should be kept in all work areas. All containers should be washed and disinfected daily; plastic liners may also be used as a cheap way to keep trash. Close-fitting lids should be kept on the containers, except when they are being filled and emptied. Containers in food-processing and food preparation areas should not be used for litter and rubbish from other areas.

Toxic Substances

Poisons and toxic chemicals should not be stored near food products. Only chemicals required for cleaning should be stored in the building, and these should be clearly labeled. Only cleaning compounds, supplies, utensils, and equipment approved by regulatory or other agencies should be used in food handling, processing, and preparation.

SUMMARY

- Food products are attractive breeding sites for microorganisms because they are rich in nutrients.
- Most microorganisms come from water, air, dust, equipment, sewage, insects, rodents, and employees.
- The chain of infection is a model that shows how foodborne diseases are carried.

- Raw materials can also be contaminated from the soil, sewage, live animals, external surfaces (skin, shells, etc.), and internal organs of meat animals. With modern health care, diseased animals rarely cause illness.
- Chemicals can contaminate foods by accidental mixing.
- Good housekeeping, sanitation, storage, and garbage-disposal practices prevent and control contamination of food.

BIBLIOGRAPHY

Bryan, F. L. 1979. Epidemiology of foodborne diseases. In *Food-Borne Infections and Intoxications,* 2d ed., p. 4–69, H. Riemann and F. L. Bryan, eds. Academic Press, New York.

Fields, M. L. 1979. *Fundamentals of Food Microbiology.* AVI Publishing Co., Westport, Conn.

Guthrie, R. K. 1988. *Food Sanitation,* 3d ed. Van Nostrand Reinhold, New York.

Hobbs, B. C., and Gilbert, R. J. 1978. *Food Poisoning and Food Hygiene,* 4th ed. Food & Nutrition Press, Westport, Conn.

Judge, M. D., Aberle, E. D., Forrest, J. C., Hedrick, H. B., and Merkel, R. A. 1989. *Principles of Meat Science.* 2d ed. Kendall Hunt Publishing Co., Dubuque, Iowa.

Lechowich, R. V. 1980. Controlling microbial contamination of animal products. Unpublished data.

Marriott, N. G. 1994. *Principles of Food Sanitation,* 3d ed. Chapman & Hall, New York.

Todd, E. C. D. 1980. Poultry-associated foodborne disease—Its occurrence, cost, sources, and prevention. *J. Food Prot.* 43:129.

STUDY QUESTIONS

1. What is the difference between foodborne infection and foodborne intoxication?
2. Give three reasons why seafood is easily contaminated.
3. What is the most common source of contamination of food?
4. List two ways that foods can be protected from microbes on hands.
5. How can food be protected against microbes in air?
6. Describe three features of buffet serving areas that protect the food from microbial contamination.
7. Where and how should garbage or trash be stored?

TO FIND OUT MORE ABOUT FOOD CONTAMINATION

1. Call the USDA Meat and Poultry Hotline, (1-800) 535-4555. Ask for information about keeping meat and poultry free from contamination.
2. Call the U.S. Environmental Protection Agency Office of Drinking Water SAFE DRINKING WATER HOTLINE, (1-800) 426-4791. Ask for information about keeping water supplies free from contamination.

3. Look in the telephone directory and call your local water utility for information on how contaminants are controlled in your local water supply.

4. Check the refrigerator and cupboards in your own kitchen. Are all foods covered? Are cleaning products stored away from food? Is the garbage can clean and covered?

MONITORING OUTBREAKS OF FOODBORNE DISEASE

Between 1983 and 1987, a total of 2,397 outbreaks of foodborne disease representing 91,678 cases were reported to the Centers for Disease Control (CDC). The cause could not be determined in 62% of outbreaks, but the following table shows the proportion of outbreaks and cases caused by various agents when the cause was known.

Cause	Outbreaks, %	Cases, %
Bacteria	66	92
Chemicals	26	2
Viruses	5	5
Parasites	4	<1

Salmonella caused 57% of the bacterial disease outbreaks and was the most common pathogen. Fish poisoning due to ciguatoxin and scomrotoxin caused 73% of the outbreaks due to chemical agents. *Trichinella spiralis* caused all parasitic disease outbreaks for 3 of the 5 years; three *Giardia* outbreaks occurred during the other 2 years. Hepatitis A caused 71% of the outbreaks due to viruses.

For each year from 1983 to 1987, the most commonly reported food preparation practice that contributed to foodborne disease was improper storage or holding temperatures, followed by poor personal hygiene of the food handler. Food obtained from an unsafe source was the least commonly reported factor. Inadequate cooking and contaminated equipment each ranked third or fourth in each of the 5 years.

Many more outbreaks occur than are reported to CDC. An outbreak is more likely to be reported if it is easily recognized and the source is confirmed by laboratory analysis of feces, blood, or food. Data on sporadic, individual cases of foodborne disease are not included, but are far more common than outbreaks.

CDC monitors foodborne disease outbreaks for three reasons:

1. *Disease prevention and control.* Surveillance of foodborne disease allows contaminated foods to be removed from stores quickly, shows when faulty food preparation practices need to be corrected in foodservice and at home, and identifies human carriers of foodborne pathogens, who can then be treated.

2. *Knowledge of causes of disease.* The pathogen causing a foodborne disease outbreak is often unidentified, either because it was not tested, the laboratory investigation was too late or incomplete, or the pathogen could not be detected even after thorough testing. Earlier and more thorough investigations and better testing methods are needed.

3. *Administrative guidance.* Data from investigations of foodbourne disease can show emerging problems and common errors in food handling. Surveillance helps to increase the awareness of food protection methods, leads to better training programs, and encourages use of available resources.

Most reports come to CDC from state and local health departments, but some also come from federal agencies such as the Food and Drug Administration (FDA), the U.S. Department of Agriculture (USDA), the U.S. Armed Forces, and occasionally from private physicians.

Source: Bean, N. H., Griffin, P. M., Goulding, J. S., and Ivey, C. B. 1990. Foodborne disease outbreaks, 5-year summary, 1983–1987. *MMWR Surveillance Summaries* 39(SS-1):15.

CHAPTER 4

Personal Hygiene and Food Handling

ABOUT THIS CHAPTER

In this chapter you will learn:

1. What personal hygiene is and why it is important
2. About parts of the body that need to be kept clean or covered
3. When and how to wash hands to keep food safe
4. How diseases are spread
5. How employees and employers can take responsibility for personal hygiene

INTRODUCTION

Food handlers can carry bacteria that can cause illness in people who eat the food handlers prepare. In fact, people are the most common source of food contamination. Hands, breath, hair, sweat, coughs, and sneezes all carry microorganisms. Even if a food handler does not feel sick, he or she could still be carrying microorganisms that can cause illness if they get into food.

The food industry spends a lot of time educating and training employees. Supervisors and workers need to understand the importance of protecting food. Foodborne disease outbreaks are very bad for business. A foodborne illness outbreak can cost about $75,000 per foodservice establishment. Investigation, cleanup, restaffing and restocking, wasted food products, settlements, and regulatory sanctions are all expensive. In multiunit restaurant chains, negative public opinion can affect sales even at units that were not involved in the outbreak.

PERSONAL HYGIENE

The word *hygiene* means using sanitary principles to maintain health. Personal hygiene refers to the cleanliness of a person's body and clothes. Food workers need to be healthy and clean to prepare safe food.

Employee Hygiene

Employees who are ill should not come to work. They should not touch food or equipment and utensils used to process, prepare, and serve food. Food can carry several illnesses, including:

- Respiratory diseases, e.g., colds, sore throats, pneumonia, scarlet fever, and tuberculosis
- Gastrointestinal diseases, e.g., vomiting, diarrhea, dysentery
- Typhoid fever
- Infectious hepatitis

After people recover from the disease, they often become carriers. This means that they still carry the disease-causing microorganisms in or on their body.

When employees are ill, they carry many more microorganisms, so they are much more likely to contaminate food. Symptoms of illnesses show that the number of microorganisms in the body is increasing, although sometimes they increase before symptoms appear. Anyone with a sinus infection, sore throat, nagging cough, or other cold symptom is probably carrying a heavy load of a virus. People who have diarrhea or an upset stomach are also probably carrying large numbers of microbes. Even after the symptoms have gone away, some of the microorganisms that caused the illness may stay in the person's body and could contaminate food. For example, an employee may carry *Salmonellae* for several months after recovering from salmonellosis. The virus that causes hepatitis may still be in the intestinal tract over 5 years after the symptoms are over. To understand why employees need good personal hygiene, it helps to look at different parts of the body that can be sources of microbial contamination. Figure 4.1 shows that bacteria live on different parts of the body and how handwashing affects the number of bacteria.

Skin

The skin constantly deposits sweat, oil, and dead cells on its outer surface. When these materials mix with dust, dirt, and grease, they form an ideal place for bacteria to grow. Therefore, bacteria from skin can contaminate food. If secretions build up and bacteria continue to grow, the skin can become itchy or irritated. Food handlers may rub or scratch the area and then transfer bacteria to food when they touch it. Contaminated food has a shorter shelf life or may cause foodborne illness. Washing

HANDS
Let's see how handwashing affects the number of bacteria present

Bacteria grow on gelatin-like food (agar) in covered, sterile, plastic plates (petri dishes).

The bacteria grow rapidly when kept at a warm temperature (98.6°F), 24 to 48 hours later, small colonies or clumps of bacteria approximately the size of a pinhead or larger can be seen in the agar.

An unwashed hand that looks clean is touched to the agar.

The plate is incubated at 98.6°F for 24 hours. The heavy growth of white colonies indicates that this hand was not very clean and that millions of bacteria were present.

After washing hands for 15 seconds with hot water and soap, bacteria are reduced in number.

Washing hands with soap and water for another 15 seconds reduces the bacteria even more.

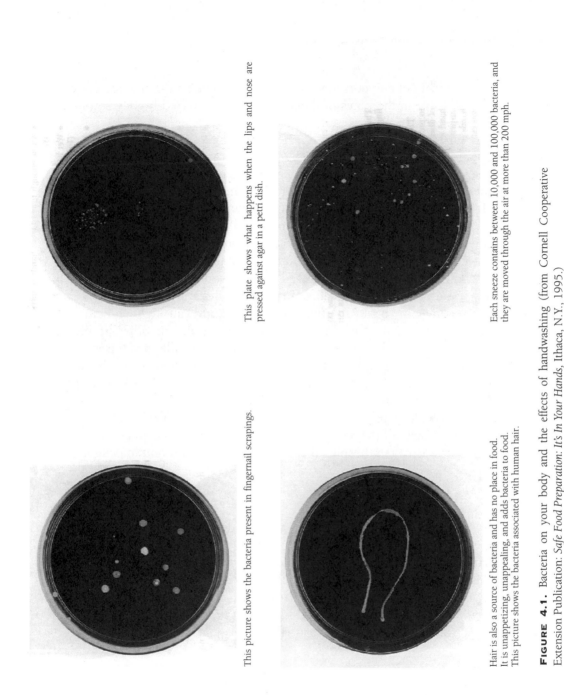

This plate shows what happens when the lips and nose are pressed against agar in a petri dish.

Each sneeze contains between 10,000 and 100,000 bacteria, and they are moved through the air at more than 200 mph.

This picture shows the bacteria present in fingernail scrapings.

Hair is also a source of bacteria and has no place in food. It is unappetizing, unappealing, and adds bacteria to food. This picture shows the bacteria associated with human hair.

FIGURE 4.1. Bacteria on your body and the effects of handwashing (from Cornell Cooperative Extension Publication: *Safe Food Preparation: It's In Your Hands*, Ithaca, N.Y., 1995.)

hands and bathing often reduces the number of microorganisms and dead cells that can be shed into the food.

Some types of bacteria do not grow on the skin, because the skin acts as a physical barrier and makes and releases chemicals that can destroy some microorganisms. This self-disinfecting system works best when the skin is clean.

The outer layer of skin (epidermis) contains cracks, crevices, and hollows where microorganisms like to live. Bacteria also grow in hair follicles and sweat glands. Because people use their hands a lot, they are often cut or callused or come in contact with a wide variety of microorganisms. Bacteria that live on the skin are not easily removed. They live in the pores of the skin and are protected by oily secretions from the glands.

Poor skin care and skin disorders do more than affect appearance. They also cause bacterial infections like boils and impetigo.

Boils are severe local infections caused by infections in hair follicles or skin glands after the outer layer of skin (epidermis) is damaged, for example by irritating clothing. Staphylococci or other microorganisms multiply in the hair follicle or skin gland and produce a toxin that kills the cells around it and causes swelling and soreness. The body collects lymph, blood cells, and tissue cells in the infected area to counteract the toxin. The body forms a barrier around the boil to prevent the infection from spreading. A boil should never be squeezed. If it is squeezed, the infection may spread and cause a cluster of boils called a *carbuncle*. If staphylococci get into the bloodstream, they may be carried to other parts of the body, causing meningitis (infection of the membranes around the brain), bone infections, or other problems. Employees with boils should be very careful if they have to handle food, because the boil is a major source of pathogenic staphylococci. Employees should use a hand dip for disinfection after touching a boil or pimple. Boils can be prevented by keeping skin clean and wearing clean clothing.

Impetigo is a very infectious skin disease caused by *Staphylococci* bacteria. The infection spreads from one part of the body to another by touching. It is common in young people who do not wash often. Keeping the skin clean helps to prevent impetigo.

Fingers

Bacteria may be picked up by the hands when they touch dirty equipment, contaminated food, clothing, or parts of the body. Employees should wash hands frequently and use a hand-dip sanitizer after touching these things so that they do not contaminate food. Plastic gloves can be used, although food can be quickly contaminated if good handwashing and other hygienic practices are not used. Plastic gloves help prevent pathogenic bacteria on the fingers and hands from reaching the food. They also help those who watch the food being prepared to feel safe while eating it.

Fingernails

One of the easiest ways to spread bacteria is through dirt under the fingernails. Employees should never handle food if their fingernails are dirty. Food handlers should not have long fingernails or artificial fingernails. Washing the hands with soap and water removes transient bacteria (bacteria that have been picked up but are not growing and multiplying), and using a hand soap that contains an antiseptic or sanitizer controls resident bacteria (bacteria that live and grow there). A humectant (moisturizer) that contains alcohol can help to control and remove both transient and resident bacteria without irritating the hands.

Jewelry

Employees should not wear jewelry in food-processing or foodservice areas. It can get caught in machinery, causing a safety hazard. Also, contaminated jewelry can fall into or come in contact with food.

Hair

Hair carries microorganisms, especially staphylococci. Employees should wear a hair-net or hat while handling food and use a hand dip after scratching their heads.

Eyes

Eyes do not normally carry bacteria. But an eye with a mild bacterial infection may have bacteria on the eyelashes and at the corners. If employees with eye infections rub their eyes, their hands will be contaminated.

Mouth

Many bacteria are found in the mouth and on the lips. It is easy to show this by asking someone to press his or her lips on the surface of sterile agar medium in a petri dish. After 2 or 3 days at room temperature, several colonies of bacteria will grow where the lips touched the agar (see Fig. 4.1). Many bacteria and viruses found in the mouth can cause disease, especially if an employee is ill. Smoking should not be allowed at work, because the smoke can carry bacteria from the mouth, and hands pick up bacteria when they carry smoking implements to the mouth. During a sneeze, some bacteria from the mouth are released into the air. The microorganisms may infect other people or may land on food while it is being handled.

Smoking, an irritating taste in the mouth, or a head cold may make an employee want to spit. Spitting should never be allowed in a food-processing operation. Spitting spreads disease, contaminates food, and does not look good. Regular tooth-brushing prevents buildup of plaque on the teeth and reduces the number of bacteria that might be carried to a food product if an employee accidentally gets saliva on his or her hands or sneezes.

Nose, Throat, and Lungs

The nose and throat have fewer bacteria than the mouth. The hairs and mucus in the upper respiratory tract filter out most of the microbes that are breathed in. The rest are destroyed by the body's defenses.

An employee who has a cold should use a hand-dip sanitizer after blowing his or her nose. Otherwise, bacteria from the nose can be transferred to the food being handled. Employees should use their elbows or shoulders to block a sneeze or cough.

A *sinus infection* means infection of the membranes in the nasal sinuses. The mucous membranes become swollen and inflamed, and secretions collect in the blocked cavities. The pressure buildup in the cavities causes pain, dizziness, and a running nose. Employees with nasal discharge should be very careful if they have to handle food products. Employees should use decongestants to reduce discharge, wash and disinfect their hands after blowing their noses, and block all sneezes with tissues.

A *sore throat* is usually caused by a type of streptococci. "Strep throat," sore throat, *laryngitis,* and *bronchitis* are spread by coughs, sneezes, and nasal discharge from carriers. Also, streptococci cause scarlet fever, rheumatic fever, and tonsillitis. These diseases spread quickly if employees practice poor hygiene.

Influenza (flu) infects the body through the lungs. Secondary bacterial infections by staphylococci, streptococci, or pneumococci can cause death.

Waste Organs

Intestinal waste (feces) is a major source of bacterial contamination. About one-third of the dry weight of stools is made up of bacteria. Particles of feces collect on the hairs around the anus and spread to underclothes. When employees use the rest room, they may get some intestinal bacteria on their hands. If they do not wash their hands properly, the microorganisms will spread to any food that they touch. Because of poor personal hygiene, the bacteria found in feces are often found in foods. Therefore, employees should wash their hands with soap before leaving the washroom, and use a hand-dip sanitizer before handling food.

The intestines of humans and animals carry common bacteria, such as *Salmonella, Shigella,* and enterococci. When these bacteria multiply, they are pathogenic (disease causing). They cause intestinal disorders and slight or severe infections, and may even cause death.

Contamination of Food Products by People

Several factors affect whether and how microbes from people contaminate food.

1. *Body site.* Different areas of the body carry different types of bacteria. The face, neck, hands, and hair are more densely populated by bacteria than are other body parts. Exposed areas of the body are more likely to be contaminated by bacteria in the air or on surfaces, or by touching other people. Hair encourages growth of microbes because of its density and oil production.
2. *Age.* The type and amount of bacteria people carry change as they grow older. These changes are most noticeable for adolescents going through puberty. Adolescents' skin glands produce large quantities of oily sebum that encourages the microorganisms that cause acne.
3. *pH.* The pH or acidity of the skin is affected by secretion of lactic acid from the sweat glands, production of fatty acids by bacteria, and diffusion of carbon dioxide through the skin. Skin usually has a pH of about 5.5. This encourages the growth of the resident microorganisms (ones that live on the skin all the time) and discourages transient microbes (ones that come and go). Soaps, creams, and other products that change the pH of the skin also alter the type of bacteria that grow.

4. *Nutrients.* Sweat contains water-soluble nutrients, and sebum contains oil-soluble nutrients. Microbiologists do not know how much these nutrients affect growth of microorganisms.

People are the biggest source of food contamination. They act as disease carriers. A carrier is a person who harbors and releases pathogens but does not have symptoms of the disease.

Handwashing

Improper handwashing causes about 25% of foodborne illnesses. Handwashing with soap and water removes bacteria. Rubbing the hands together or using a scrub brush removes more bacteria than quick handwashing.

Antimicrobial agents remove more bacteria than ordinary hand soap, but employees need to use an antimicrobial hand soap throughout the day for it to be fully effective. The antimicrobial agent needs to be in contact with the hands for more than 5 seconds to have an effect on the number of microbes. Figure 4.2 illustrates the recommended double handwashing procedure.

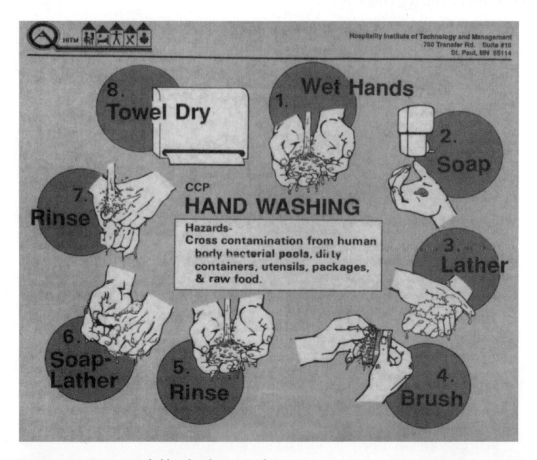

FIGURE 4.2. Recommended handwashing procedure.

Methods of Disease Transfer

Direct. Many disease microorganisms are transferred directly from one person to another through close contact. Examples are diphtheria, scarlet fever, influenza, pneumonia, smallpox, tuberculosis, typhoid fever, dysentery, and venereal diseases. Respiratory diseases are transferred via tiny particles released from the nose and mouth when a person talks, sneezes, or coughs. The particles can be suspended in the air, attached to dust, for a long time. Other people can be infected when they inhale the particles.

Indirect. The host (infected person or animal) of an infectious disease may transfer organisms to water, food, objects, or soil. Doorknobs, telephones, pencils, books, washroom fittings, clothing, money, and knives can all transfer infections from one user to another. Salmonellosis, dysentery, and diphtheria are examples of diseases that can be spread indirectly. To reduce indirect transfer of microorganisms, washbasins should have foot-operated controls instead of hand-operated faucets, and doors should be self-closing.

Requirements for Hygiene

Management must have a protocol to make sure employees use hygienic practices. Supervisors and managers should set an example for employees by using excellent hygiene and health practices themselves. They should provide proper laundry, lockerroom, and handwashing facilities to make it easy for employees to stay clean and hygienic.

All employees should have a physical examination before they are employed to check that they have good physical, mental, and emotional health. All employees who work with food should be checked regularly for signs of illness and infection, and other signs of poor health.

Employees should maintain personal hygiene in the following ways. They should:

1. Maintain good physical health through good nutrition, enough rest, and physical cleanliness
2. Report illness to their employer before working with food so that assignments can be adjusted to protect food from being contaminated
3. Practice good hygiene so that they do not contaminate food
4. Wash their hands during their work shift after using the toilet; after handling garbage or other dirty items; after handling uncooked meats, egg products, or dairy products; after handling money; after smoking; after coughing and sneezing; and when leaving or returning to food production/service areas
5. Maintain personal cleanliness through daily bathing, washing hair at least twice a week, cleaning fingernails daily, use of a hat or hairnet while handling food, and wearing clean underclothing and uniforms
6. Not touch foodservice equipment and utensils with their hands, and use disposable gloves if they have to touch food (other than dough)

7. Follow rules, such as "no smoking," and do anything else needed to protect the food from being contaminated

Employers should help employees use good hygiene in the following ways. Employers should:

1. Train employees in food handling and personal hygiene.
2. Watch and monitor employees and their work habits regularly. Employers should treat poor hygiene practices as disciplinary violations.
3. Give incentives to encourage excellent hygiene and sanitary practices.

Food handlers are responsible for their own health and personal cleanliness. Employers are responsible for making sure that the public is protected from unsanitary practices that could cause illness. Personal hygiene is essential for production of wholesome food.

HYGIENIC FOOD HANDLING

Food handlers should protect food by using barriers to separate the food from anything that could contaminate it at all stages of production. Barriers include disposable gloves, hairnets, mouth guards, sneeze guards, lids, and food wraps.

Hiring Employees

Careful hiring helps the company keep a good image and meet regulations. Over 50% of people carry *Staphylococcus aureus* in their mouths and nasal passages. The American Disease Act prohibits discrimination in hiring on the basis of disease, especially human immunodeficiency virus (HIV) status. Many carriers of disease do not have obvious symptoms. Many areas no longer require health cards because they are expensive to get, and workers can be infected after they get their card. Employees should be hired only if they meet the following conditions:

1. Applicants should not have obvious hazards, such as open sores or excessive skin infections, pimples, or acne.
2. Applicants should not have respiratory problems if they will be handling food or working in food-processing or food-preparation areas.
3. Applicants and their clothing should be neat, clean, and fresh smelling.
4. Applicants should pass a sanitation course and examination, e.g., one provided by the National Restaurant Association.

Personal Hygiene Rules

Food organizations should have clear and strict personal hygiene rules. The rules should be clearly posted on the wall or spelled out in booklets given to each employee. The policies should cover personal cleanliness, clothing, good food-handling practices, and use of tobacco.

Facilities

Employees need the right equipment and supplies to maintain good personal hygiene. Locker rooms and rest rooms should be clean, neat, well lighted, and conveniently located away from production areas. Rest rooms should have self-closing doors. Ideally, faucets at handwashing stations should be foot- or knee-operated and supply water at 43 to 50°C (109–122°F). Liquid soap dispensers are better than bars of soap, because bars can transfer microorganisms from one person to the next. Disposable towels are best for drying hands. Food and beverages should be consumed in a special area. This area should be clean and pest-free.

Supervising employees. Employees who handle food should follow the same health standards used to screen new employees. Supervisors should watch employees daily for infected cuts, boils, respiratory diseases, or other infections. Many local health departments require foodservice and food-processing firms to report employees who may have or carry an infectious disease.

Employee responsibilities. The employer is responsible for the way employees handle food, but employers should make employees responsible for the following activities when they start work. Employees should:

1. Not have respiratory, gastrointestinal, or other diseases
2. Report injuries—including cuts, burns, boils, and skin problems—to their employer
3. Report respiratory illnesses (e.g., colds, influenza, sinus infections, and bronchitis) and intestinal illnesses, such as diarrhea, to their employer
4. Bathe daily, wash their hair at least twice a week, change their underclothes daily, and keep their fingernails clean
5. Tell their supervisor when soap, towels, or other washroom supplies run out
6. Break such habits as scratching their heads or touching other parts of their body
7. Cover their mouths and noses when they cough or sneeze
8. Wash their hands after using the toilet, blowing their nose, smoking, handling anything dirty, and handling money
9. Keep their hands out of food. Food should not be tasted using a hand and should not be eaten in food production areas
10. Not use utensils that touch their mouths to handle food
11. Use disposable gloves to handle food
12. Not smoke or chew tobacco in food production and food preparation areas and should wash their hands after smoking

SUMMARY

- Microorganisms that cause illness and food spoilage often come from food handlers.
- *Hygiene* means using sanitary practices to maintain good health.

- Personal hygiene means maintaining a clean body and clean clothing.
- Parts of the body and clothing that can contaminate food include the skin, hands, fingernails, jewelry, hair, eyes, mouth, nose and lungs, and waste organs. Employees may carry a disease-causing microorganism even if they do not have symptoms of the illness.
- Microorganisms can be transferred directly (from person to person) or indirectly (via food, water, equipment, and surfaces).
- Management must hire employees who are healthy, clean, and neat. They must train employees to use good hygiene. Managers must hold employees responsible for personal hygiene so that the food that they handle is safe to eat.

BIBLIOGRAPHY

Anon. 1976. *Plant Sanitation for the Meat Packing Industry.* Office of Continuing Education, University of Guelph and Meat Packers Council of Canada.

Hobbs, B. C., and Gilbert, R. J. 1978. *Food Poisoning and Food Hygiene,* 4th ed. Food & Nutrition Press, Westport, Conn.

Longree, K., and Blaker, G. G. 1982. *Sanitary Techniques in Food Service.* John Wiley, New York.

Marriott, N. G. 1994. *Principles of Food Sanitation,* 3d ed. Chapman & Hall, New York.

National Restaurant Association Education Foundation. 1991. *Applied Foodservice Sanitation,* 4th ed. John Wiley, in cooperation with the Education Foundation of the National Restaurant Association, Chicago.

Wise, H. S. 1979. The food handler and personal hygiene. In *Sanitation Notebook for the Seafood Industry,* p. II-1, G. J. Flick, Jr., et al., eds. Department of Food Science and Technology, Virginia Polytechnic Institute and State University, Blacksburg.

STUDY QUESTIONS

1. What should food handlers do at home to maintain good personal hygiene?
2. When should food handlers wash their hands while at work?
3. Why do healthy employees need to be careful not to contaminate food?
4. List seven body parts that are often sources of microbial infection.
5. Give two reasons why employees should not wear jewelry.
6. How can employers help employees maintain good personal hygiene?

TO FIND OUT MORE ABOUT PERSONAL HYGIENE

1. Observe a food handler (or ask someone else to watch you). Note how often they touch their face or hair, blow their nose, sneeze, and cough. Do they block their sneezes and coughs? Do they wash their hands afterwards?
2. Check the handwashing facilities at work, school, or home. Are they close to food preparation areas? Are they clean? Do they have hot and cold water, liquid soap, a nail brush, and hand towels or blow drier?

3. Put scrapings from under your fingernails, or press your lips on a sterile agar plate. Cover the plate and keep it at room temperature for 2 to 3 days. How may colonies of bacteria have grown?

4. Contact the supplier or manufacturer of the antimicrobial soap used at your workplace or school. Ask for information about good hand washing and personal hygiene.

5. Obtain a Glo-Germ kit from the Glo Germ Company, P.O. Box 537, Moab, UT 84532, (1-800) 842-6622. This kit contains an oil, a powder, and a special fluorescent lamp. The oil and powder contain plastic germs. You can put either the oil or the powder on your hands and then wash your hands normally. The fluorescent lamp shows up any remaining germs and shows how effectively you have washed your hands.

GOOD HANDWASHING *Is* IMPORTANT

These three reports show how easily poor handwashing can cause major outbreaks of foodborne illness.

On a 4-day Caribbean cruise, 72 passengers and 12 crew members had diarrhea, and 13 people had to be hospitalized. Stool samples of 19 of the passengers and 2 of the crew members contained *Shigella flexneri* bacteria. The illness was traced to German potato salad prepared by a crew member who carried these bacteria. The disease spread easily because the toilet facilities for the galley crew were limited (1).

Over 3,000 women who attended a 5-day outdoor music festival in Michigan became ill with gastroenteritis caused by *Shigella sonnei*. The illness began 2 days after the festival ended, and patients were spread all over the United States before the outbreak was recognized. An uncooked tofu salad served on the last day caused the outbreak. Over 2,000 volunteer food handlers prepared the communal meals served during the festival. Before the festival, the staff had a smaller outbreak of shigellosis. Sanitation at the festival was mostly good, but access to soap and running water for handwashing was limited. Good handwashing facilities could have prevented this explosive outbreak of foodborne illness.

Shigella sonnei caused an outbreak of foodborne illness in 240 airline passengers on 219 flights to 24 states, the District of Columbia, and 4 countries. The outbreak was identified only because it involved 21 of 65 football team players and staff. Football players and staff, airline passengers, and flight attendants with the illness all had the same strain of *S. Sonnei*. The illness was caused by cold food items served on the flights that had been prepared by hand at the airline flight kitchen. Flight kitchens should minimize hand contact when preparing cold foods or remove these foods from in-flight menus.

Sources:

(1) Lew, J. F., Swerdlow, D. L., Dance, M. E., et al. 1991. An outbreak of shigellosis aboard a cruise ship caused by a multiple-antibiotic-resistant strain of *Shigella flexneri*. *Am. J. Epidemiol.* 134: 413.

(2) Lee, L. A., Ostroff, S. M., McGee, H. B., et al. 1991. An outbreak of shigellosis at an outdoor music festival. *Am. J. Epidemiol.* 133: 608.

(3) Hedberg, C. W., Levine, W. C., White, K. E., et al. 1992. An international foodborne outbreak of shigellosis associated with a commercial airline. *J. Am. Med Assoc.* 268: 3208.

Cleaning Compounds

ABOUT THIS CHAPTER

In this chapter you will learn:

1. About different types of soil
2. How soil is attached to and can be removed from surfaces
3. How cleaning compounds work
4. How and when to use different cleaning and scouring compounds
5. How to choose a cleaning compound
6. How to handle and use cleaning compounds safely
7. How cleaning chemicals can contaminate food

INTRODUCTION

Cleaners are made for specific jobs, such as washing floors and walls or use in high-pressure dishwashers. Good cleaners are economical, easy to measure, and dissolve well. They are approved for use on food surfaces, are not corrosive, and do not cake, leave dust, or break down during storage.

Different cleaning compounds work well for different areas and different types of equipment. When choosing a cleaning compound, it is important to consider the type of soil (dirt), the water supply, how the cleaner will be used, and the area and kind of equipment being cleaned.

TYPES OF SOIL

Soil is material in the wrong place. It is made up of dirt, dust, and scraps of food. Examples of soil are fat smears on a cutting board, lubricant on a moving conveyor belt, and food scraps on processing equipment.

It is important to select the correct cleaning compound to remove a specific type of soil. Table 5.1 shows whether various kinds of soil are soluble in water, acid, or alkali; whether heat helps to remove them; and how hard they are to remove. An acid cleaning compound works best to remove inorganic deposits; an alkaline cleaner is more effective for removing organic deposits. Table 5.2 shows examples of different types of inorganic and organic soil.

TABLE 5.1.   Removing Different Types of Soil

Type of Soil	Solubility	Ease of Removal	Effects of Heat
Salts	Soluble in water Soluble in acid	Easy to difficult	Reacts with other types of soil and becomes harder to remove
Sugar	Soluble in water	Easy	Caramelizes and becomes difficult to remove
Fat	Insoluble in water Soluble in alkali	Difficult	Molecules join together and become difficult to remove
Protein	Insoluble in water Slightly soluble in acid Soluble in alkali	Very difficult	Molecules change shape (denature) and become very difficult to remove

TABLE 5.2.   Types of Soil

Type of Soil	Soil Subclass	Examples
Inorganic soils	Hard-water deposits	Calcium and magnesium carbonates
	Metallic deposits	Common rust, other oxides
	Alkaline deposits	Films left when an alkaline cleaner is not rinsed off properly
Organic soils	Food deposits	Food scraps and specs
	Petroleum deposits	Lubrication oils, grease, and other lubricants
	Nonpetroleum deposits	Animal fats and vegetable oils

Soil deposits are often complex mixtures of organic and inorganic materials. It is important to know the type of soil and use the best cleaning compound or combination of compounds to remove it. Employees often need to use a two-step cleaning procedure, using more than one cleaning compound to remove a combination of inorganic and organic deposits. Table 5.3 shows the best type of cleaning compound for each type of soil.

TABLE 5.3.   Cleaning Compounds for Different Types of Soil

Type of Soil	Cleaning Compound
Inorganic soils	Acid cleaner
Organic soil	
Nonpetroleum	Alkaline cleaner
Petroleum	Solvent cleaner

HOW SOIL IS ATTACHED

Type of Surface

Something else to consider when choosing a cleaner is the type of surface (see Table 5.4). The type of surface affects the type of soil that collects and how it is removed.

Sanitation employees should know about the finishes on all equipment and areas in the food facility and which cleaning chemicals could attack each surface. If management is not familiar with the cleaning compounds and surface finishes, they should ask a consultant or reputable supplier of cleaning compounds for technical assistance.

Removing Soil from Surfaces

Soil is difficult to remove from cracks, crevices, and other uneven surfaces, especially in hard-to-reach areas. It is easiest to remove soil from surfaces that are smooth, hard, and nonporous. Removal of soil from a surface takes three steps: (1) separating the soil from the surface, (2) dispersing the soil in the cleaning solution, and (3) preventing dispersed soil from reattaching to the surface.

Step 1: *Separating soil from the surface, material, or equipment to be cleaned.* Soil can be separated mechanically (using high-pressure water, steam, air, or scrubbing) or chemically (for example, the reaction of an alkali cleaner with a fatty acid to form a soap).

The soil and surface must be thoroughly wet for a cleaning compound to help separate the soil from the surface. The cleaning compound reduces the strength of the bond between the soil and the surface, so that the soil is loosened and separated. Heat or mechanical action (scrubbing, shaking, or high-pressure spray) can help reduce the strength of the bond (although heat does not help loosen some protein and fat soils).

Step 2: *Dispersing soil in the cleaning solution.* Dispersion means diluting soil in a cleaning solution. Cleaning staff must use enough cleaning solution to dissolve all of the soil. More soil will need more cleaning solution. As the cleaning solution becomes saturated with soil, staff need to use fresh cleaning solution.

Some soils that have been loosened from the surface will not dissolve in the cleaning solution. Dispersion of soils that do not dissolve is more complicated. It is important to break up the soil into smaller particles or droplets that can be carried

TABLE 5.4. Characteristics of Surfaces in Food-Processing Plants

Material	Characteristics	Precautions
Wood	Soaks up moisture, fats, and oils. Difficult to maintain. Softened by alkali. Destroyed by caustics.	Do not use, because it is unsanitary. Use stainless steel, polyethylene, or rubber instead.
Black metals	Acid or chlorinated detergents may cause rust.	Often tinned or galvanized to prevent rust. Use neutral detergents to clean.
Tin	May be corroded by strong alkaline or acid cleaners.	Do not allow tin surfaces to touch foods.
Concrete	May be etched by acid foods and cleaning compounds.	Concrete should be dense and acid resistant. Should not make dust. Can use acid brick instead.
Glass	Smooth and impervious. May be etched by strong alkaline cleaning compounds.	Clean glass with moderately alkaline or neutral detergents.
Paint	Method of application affects surface quality. Etched by strong alkaline cleaning compounds.	Some edible paints can be used in food plants.
Rubber	Should not be porous or spongy. Not affected by alkaline detergents. Attacked by organic solvents and strong acids.	Rubber cutting boards can warp and their surface dulls knife blades.
Stainless steel	Generally resists corrosion. Smooth, impervious surface. Resists oxidation at high temperatures. Easy to clean. Nonmagnetic.	Stainless steel is expensive and may not be readily available in the future. Some stainless steel is attacked by halogens (chlorine, iodine, bromine, and fluorine).

away from the cleaned surface. Mechanical energy (shaking, high-pressure water, or scrubbing) helps cleaning compounds break down the soil into small particles.

Step 3: *Preventing dispersed soil from reattaching to the surface.* Cleaning methods should:

- Remove the dispersed solution from the surface being cleaned.
- Shake the dispersed solution to stop the soil from settling.
- Prevent reactions between the cleaning compound and water on the soil (i.e., by using soft water to prevent deposits formed when hard water reacts with soap in the cleaning compound).

- Flush or rinse the cleaned surface to remove all dispersed soil and cleaning solution residues.
- Keep the soil finely dispersed so that it does not become trapped on the cleaned surface.

Successful soil removal depends on several factors:

- Cleaning procedures
- Proper supervision
- Cleaning compounds
- Water quality
- High-pressure application of cleaning solutions
- Mechanical shaking
- Temperature of the cleaning compounds and solutions

How Cleaning Compounds Work

Food particles and other debris provide nutrients for microorganisms to grow. Food particles also protect microorganisms during cleaning by neutralizing the effects of chlorinated cleaning compounds and sanitizers so that they cannot get to the microbes. Cleaning staff must completely remove all soil using mechanical energy and cleaning compounds before they can inactivate the microbes using sanitizers.

Two Ways They Work

Cleaning compounds work in two ways:

- They lower the energy of the bond between the soil and the surface so that the soil can be dislodged and loosened.
- They suspend soil particles in the solution so that they can be flushed away.

To complete the cleaning process, a sanitizer is applied to destroy residual microorganisms that are exposed through cleaning.

Good-quality water helps cleaners to work. The water should be free of minerals (i.e., soft water), free of microorganisms, clear, colorless, and noncorrosive. Hard water contains minerals that may react with some cleaning compounds and prevent them from working properly (although some cleaning compounds can overcome the effects of hard water).

One of the oldest and best-known cleaning compounds is plain soap. Fats, oils, and grease do not dissolve in water, but soap disperses tiny particles of these materials in the solution. After the soap disperses the fat or oil, the soil is easily flushed away. Dispersion of materials that are insoluble in water by soap is called *emulsification*. In emulsification, the cleaning compound molecules surround soil particles and suspend them in the cleaning solution (see Fig. 5.1). Food-processing and foodservice operations rarely use soap because it does not clean well and reacts with hard water to form an insoluble curd (like the ring around the bathtub). Many food operations use detergents instead of soaps.

Cleaning-Compound Terms

Sanitation workers need to understand the terms used to describe cleaning compounds. Each manufacturer uses its own brand names and codes. This text provides generic information, rather than endorsing brand names.

Chelating agent (sequestering agent or sequestrant): Chemical added to cleaning compounds to prevent the salts of calcium and magnesium in hard water from forming deposits on equipment surfaces (i.e., scale).

Emulsification: Breakdown of fat and oil drops into smaller droplets that are dispersed in the cleaning solution. The soil is still there, but the particles are smaller and are dispersed in the solution, rather than settling on the surface.

Rinsibility: The ability of a cleaning compound to be removed from a surface without leaving a residue.

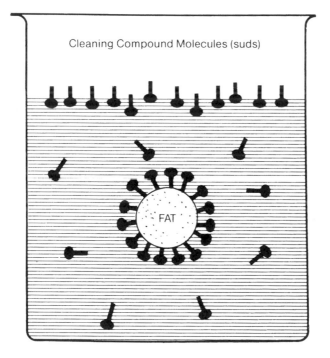

FIGURE 5.1. Suspended soil particle.

Surfactant: A complex molecule that is blended with a cleaning compound to reduce the energy of the bonds around the soil and allow closer contact between the soil and the cleaning compound.

Suspension: The process of loosening, lifting, and holding soil particles in solution.

Water hardness: The amount of inorganic salts (such as calcium chloride, magnesium chloride, sulfates, and bicarbonates) in water.

Water softening: Removes or inactivates the calcium and magnesium ions in water.

Wetting (penetration): Caused by a surfactant that allows the cleaning compound to wet or penetrate the soil deposit and loosen it from the surface.

TYPES OF CLEANING COMPOUNDS

Most cleaning compounds used in the food industry are blended products. Manufacturers combine ingredients to make a specific product for a particular type of surface or dirt. The following types of cleaning compounds are used most often in foodservice facilities and processing plants. You will see several of these chemicals in a typical cleaning compound.

Alkaline Cleaning Compounds

Alkaline cleaning solutions have a pH between 7 (neutral) and 14 (most alkaline). There are several types of alkaline cleaners.

Strongly alkaline cleaners. These cleaners have strong dissolving powers and are very corrosive. If these cleaners come in contact with skin they can cause burns, ulcers, and scarring; prolonged contact may cause permanent damage. Inhaling the fumes or mist damages the lungs.

An example of a strongly alkaline compound is sodium hydroxide (caustic soda), which destroys microbes, dissolves protein, and is good at dispersing and emulsifying soil. Silicates make sodium hydroxide less corrosive, better at penetrating soil, and better at rinsing away soil. These cleaners are used to remove heavy soils, such as those in commercial ovens and smokehouses, but they are not good at removing mineral deposits. They are not used as manual cleaners because they can be harmful to people and equipment.

Heavy-Duty Alkaline Cleaners

These compounds have moderate dissolving powers and are either slightly corrosive or not corrosive at all. However, if they are in contact with the skin for long, they may remove necessary oils from the skin, leaving it open to infections.

These cleaners are often used for cleaning in place or high-pressure or other mechanized systems. They are very good at removing fats but do not remove mineral deposits. Sodium carbonate is quite low in cost, is widely used in heavy-duty and manual cleaning procedures, and is used to buffer many cleaning compounds. Sulfites reduce the corrosive effect of heavy-duty alkaline cleaners on tin and tinned metals.

Mild Alkaline Cleaners

Mild alkaline cleaning solutions such as sodium bicarbonate are used to clean lightly soiled areas by hand. These compounds are good at softening water but do not remove mineral deposits.

Acid Cleaning Compounds

Acid cleaning compounds remove materials that are dried on or encrusted on surfaces and dissolve mineral scale. They are especially good at removing mineral deposits formed by alkaline cleaning compounds. When hard water is heated above 80°C (176°F), some of the minerals are deposited. These deposits stick to metal surfaces and leave a rusty or whitish scale. Acid cleaners dissolve the minerals in the deposits so that they can be easily removed.

Organic acids (such as citric, tartaric, sulfamic, and gluconic acids) are also excellent water softeners, rinse off easily, and do not corrode surfaces or irritate the skin. Inorganic acids are excellent at removing and controlling mineral deposits, but they can be very corrosive to surfaces and irritating to the skin. Acid cleaning compounds are used for special purposes rather than for all-purpose cleaning. Acid cleaning compounds are less effective than alkaline compounds against the soil caused by fats, oils, and proteins.

Strongly Acid Cleaners

These compounds corrode concrete, most metals, and fabrics. Heating some acid cleaners produces corrosive, toxic gases, which can damage the lungs. Strongly acid cleaners remove encrusted surface matter and mineral scale from steam equipment, boilers, and some food-processing equipment. When the solution is too hot, the mineral scale may redeposit and form a tarnish or whitish film on the equipment being cleaned.

Phosphoric acid and hydrofluoric acid both clean and brighten certain metals. But hydrofluoric acid is corrosive to stainless steel and dangerous to handle because it tends to release hydrogen gas. Phosphoric acid is widely used in the United States. It is not very corrosive and works well with many surfactants.

Mildly Acid Cleaners

These compounds are slightly corrosive and may cause sensitivity reactions. Some acid cleaners attack skin and eyes. Examples of mildly acid cleaning compounds are levulinic, hydroxyacetic, acetic, and gluconic acids. These cleaners may contain other chemicals, such as wetting agents and corrosion inhibitors. Organic acids are good manual cleaners, are more expensive than the other acid cleaning compounds, and can soften water.

Solvent Cleaners

Solvent cleaners are based on ether or alcohol. They work well on soils caused by petroleum products, such as lubricating oils and greases. Most of the time, food establishments use alkaline cleaners to remove organic soils. But they use solvent cleaners to remove large amounts of petroleum deposits in areas free of protein-based and greasy soils, i.e., in the maintenance shop and on motors, gear boxes, pallet trucks, and fork trucks.

Solvent cleaners may be mixed with wetting agents, water softeners, and other additives. Heavy-duty solvent cleaners do not mix with water and often form an emulsion when water is added.

Soaps and Detergents

Soaps and detergents emulsify fats, oils, and grease so that they are easily washed away. Soaps and detergents usually contain chemical builders to make them clean more effectively. Chemical builders are usually alkaline. Alkalis and alkaline chemicals are sometimes called *caustics,* but the general term *bases* is more correct. Soaps and detergents for household cleaning have a pH of 8 to 9.5. Regular and prolonged use of these cleaners removes grease from the skin, but ordinary use is safe. Detergents either remove natural oils from the skin or react with the oils so that the skin becomes sensitive to chemicals that do not usually affect it. Some hand soaps are slightly acid (pH 6, the same as skin) and can clean very dirty skin without removing the grease.

ADDITIVES TO CLEANING COMPOUNDS

Manufacturers add various chemicals to cleaning compounds to protect sensitive surfaces or improve the cleaning properties of the compound.

Sequestrants

These additives are also known as *chelating agents* and *sequestering agents.* They soften water by forming complex molecules with the magnesium and calcium in hard water.

Organic chelating agents act as water conditioners and are more efficient than phosphates in sequestering calcium and magnesium ions and preventing scale buildup. These chelating agents do not break down at temperatures above 60°C (140°F) and when the solution is stored for long periods.

Surfactants

These agents help to spread cleaning and sanitizing compounds over the surface to be cleaned. Surfactants wet, penetrate, emulsify, disperse, and suspend soil particles.

Cleaning detergents are made up of a surfactant and a builder. *Builders* make cleaners more effective by controlling properties that make the surfactant less effective. Phosphates are excellent builders, especially for heavy-duty cleaning compounds.

SCOURING COMPOUNDS

Scouring compounds or chemical abrasives are normally made from neutral or mildly alkaline ingredients. Abrasives are usually mixed with soaps and used with brushes or metal sponges. Neutral scouring compounds are often mixed with acid cleaners to remove alkaline deposits and encrusted materials. Abrasive cleaning compounds should be used carefully on stainless steel to avoid scratching it.

Slightly Alkaline Scouring Compounds

Borax and sodium bicarbonate are mildly alkaline scouring compounds used for light deposits of soil. They are not good detergents or emulsifiers.

Neutral Scouring Compounds

These compounds are made from volcanic ash, pumice, silica flours, or feldspar. Cleaning powders or pastes used for manual scrubbing and scouring may contain these.

CHOOSING A CLEANING COMPOUND

It is important to choose the right cleaning compound for the type of soil. A good rule to remember is that *like cleans like*. Therefore an acid soil requires an acid cleaner, while an alkaline cleaning compound works best to remove an alkaline soil. Alkaline, general-purpose cleaning compounds work best to remove organic soils. Heavy-duty alkaline cleaning compounds work best for heavy deposits of fats and proteins (organic soil). Acid cleaning compounds remove mineral deposits (inorganic soil) and other soils that are not removed by alkaline cleaning compounds. Phosphates complexed with organic chlorine are the most common types of cleaner-sanitizers.

Type and Amount of Soil

Heavy soil may need different cleaning compounds and additives from light soil for effective cleaning. Different cleaning compounds and additives work best to remove different kinds of soil. It is important to use the right compounds so that cleaning is effective. Some types of soil need more than one cleaning compound to remove them thoroughly. It is often safer to use two cleaning compounds one after the other, because cleaning compounds can react together dangerously if they are mixed.

Temperature and Concentration

Hotter and more concentrated (stronger) cleaning-compound solutions are more active than cooler and less-concentrated solutions. However, high temperatures (above 55°C [131°F]) and stronger concentrations than the manufacturer or supplier recommends can change the shape of the protein in the soil deposits and make it harder to remove. High temperatures can also cause more etching (scratching) of surfaces.

Cleaning Time

The surface becomes cleaner when the cleaning compound is in direct contact with the soil for longer. The way the cleaning compound is used and the type of cleaner affect how long the surface is exposed to the cleaner.

Mechanical Force

Mechanical energy (shaking and high-pressure sprays) helps the cleaning compound to penetrate the soil and separate soil from the surface. Chapter 7 discusses how mechanical energy (cleaning equipment) helps remove soil.

HANDLING AND STORING CLEANING COMPOUNDS

Most cleaners absorb moisture if they are left uncovered. Containers must be kept sealed to prevent contamination and to keep the materials from caking.

Cleaning compounds should be stored away from normal traffic. This special area should have pallets, skids, or storage racks to keep the containers off the floor. The floor and air should be dry, and the temperature should be moderate so that liquid products do not freeze. The storage area should be locked to prevent theft.

An inventory sheet helps keep track of when new supplies need to be ordered and shows when employees have used unusually large or small amounts of products. One person should be responsible for controlling and supplying cleaning materials to make sure that products are not wasted and that cleaning supplies do not run out. This worker should be familiar with each cleaning operation so that he or she can teach other employees how to use various types of cleaner and cleaning equipment.

Choosing the right cleaning compound and how to use it can be complicated. Suppliers of cleaning compounds can help managers decide what compound to use and how to use it. Managers should give employees clear instructions on how to use the product so that they can clean surfaces and equipment properly without damaging them. Managers should read the instructions suppliers send with commercial cleaning compounds. They should not mix compounds from different suppliers.

Different types of soil in food plants need different mixtures of cleaning compounds. Managers of large plants buy basic cleaning compounds and blend them in batches. Staff may put together between 12 and 15 mixtures for specific cleaning jobs around the plant. Managers of smaller facilities usually buy ready-mixed formulas.

Employees should always be extremely careful when they are using cleaning compounds. Strong chemical cleaners can cause burns, poisoning, dermatitis (inflammation of the skin), and other problems.

Hazards of Alkali Cleaners

Strong alkaline cleaning compounds corrode body tissues, especially the eyes. Solid and solution forms of these cleaners are equally dangerous. The irritation begins as soon as the cleaner comes in contact with the body. These cleaners can cause burns and scarring. Even dilute solutions can destroy body tissues if the tissue is exposed for a long time. Dilute solutions can gradually remove grease from the skin, leaving it exposed and open to bacteria. Dry powder or particles can get inside gloves or shoes and cause severe burns. Inhaling the dust or a concentrated mist of alkaline solutions can damage the airways and lungs.

Many alkaline chemicals react violently when they are mixed with water. The heat released when they are mixed can make the solution boil. This releases large amounts of hazardous mist and vapor.

Hazards of Acid Cleaners

Sulfamic acid. This acid cleaner is fairly safe to use and store. It is important to store it away from heat and potential fire hazards because it produces toxic oxides of sulfur when it gets hot.

Acetic acid. This acid damages skin and eyes. It is important to store it in an area designed for flammable materials because it catches fire more easily than other acid cleaners.

Citric acid. This cleaning acid is fairly safe, although it often causes allergic reactions if it is in contact with skin for a long time. It is unlikely to catch fire, but it produces irritating fumes when it gets hot.

Hydrochloric acid (muriatic acid). This acid can easily injure someone if it is used incorrectly. Air containing 35 parts per million (ppm) hydrochloric acid vapor quickly irritates the throat. The amount of hydrochloric acid vapor in air cannot legally be higher than 5 ppm throughout an 8-hour work shift. Cleaners used to descale metal equipment often contain this acid because it reacts with tin, zinc, and galvanized coatings. Hydrochloric acid loosens the outer layers of soil and stain and carries them away. Hydrochloric acid makes the surface of concrete floors rougher and more slip resistant. If this acid gets hot or comes into contact with hot water or steam, it produces toxic and corrosive hydrogen chloride gas.

Sodium acid sulfate and sodium acid phosphate. These cleaners irritate skin and cause chemical burns if they are in contact with the skin for a long time. Solutions of these compounds are very acidic and can damage the eyes if they are not flushed immediately.

Phosphoric acid. Metal cleaners and metal brighteners contain this acid. Concentrated phosphoric acid corrodes skin and eyes very quickly. When heated for a long time, phosphoric acid produces toxic fumes of phosphorus oxides. Metal cleaners containing phosphoric acid mixed with other chemicals should be used sparingly.

Hydrofluoric acid. Hydrofluoric acid helps to clean and brighten metal, e.g., aluminum. On its own, hydrofluoric acid irritates and corrodes the skin and nasal passages. Breathing the vapor can cause ulcers in the airways and lungs. Hydrofluoric acid should be used carefully, even when it is very dilute. When it gets hot, it produces a very corrosive fluoride vapor, and it reacts with steam to produce a toxic and corrosive mist. Usually, very small amounts are used, because larger quantities can release hydrogen gas if it is in contact with metal containers. It must be stored in a safe environment like that used for flammable liquids.

Acid cleaners tend to attack the skin and eyes more slowly than alkaline cleaners. People may not immediately realize how badly they are injured. Hydrofluoric acid can destroy the oil barrier of the skin so that washing and flushing the area may not help much. Hydrofluoric acid is especially hazardous because it gives little warning of an injury until the damage is serious. It is important not to confuse hydrofluoric acid with other acids because it acts differently and needs different medical treatment.

Protective Equipment

Sanitation workers should wear waterproof, knee-high footwear to keep their feet dry. Trouser legs should be worn outside boots to stop powders, hot water, or strong cleaning solutions from getting into boots. Trouser legs can only be tucked in if the worker is wearing strap-top boots.

Employees need different types of protective equipment, depending on the strength of the solution and how it is being used. Employees should wear protective hoods, long gloves, and long aprons when cleaning solutions are dispensed from overhead sprayers. If the chemicals produce mists or gases when they are mixed or used, employees need to wear protective devices to protect their lungs and airways. Supervisors should make sure that employees wear the right size and type of equipment and that it is used and maintained properly.

Employees should wear chemical goggles or safety glasses, even when handling mild cleaning compounds. Even a mild hand soap can irritate the eyes because of the alkaline pH (pH ~9). Constant contact with mild cleaning solutions can cause dermatitis (inflamed skin). A person wearing contact lenses should wear chemical goggles when he or she works in any area where dangerous chemicals are handled.

Mixing and Using

Employees must wear an apron, goggles, rubber gloves, and dust mask when mixing dry ingredients. Only experienced, well-trained staff should prepare cleaning chemicals. The sanitation supervisor should understand the basic chemistry of the ingredients and should teach workers what they need to know to prevent accidents. Workers should know the hazards of each chemical and how chemicals react when mixed. Supervisors should make sure that workers understand that cleaning chemicals are not just soaps, and that they need protective equipment to protect them from strong and sometimes dangerous chemicals. Employees must clean protective equipment after they use it.

Employees should use cold water to mix most cleaning solutions. Only a few chemicals require hot water to dissolve. It is important to keep solutions cool during mixing so that they do not give off dangerous fumes.

Sanitarians should use the recommended concentration for each chemical. After mixing, staff should store chemicals in a clearly marked container that gives the common name of the compound, its ingredients, any precautions for use, and its concentration. Supervisors must make sure that workers do not mix stronger solutions than are recommended, thinking that if a little is good, a lot would be better. Stronger solutions may be unsafe. Workers need to understand how important it is not to mix cleaners once they are prepared. Supervisors should warn workers not to put dry chemicals back in barrels and not to mix them with unknown chemicals.

Storing and Transporting Chemicals

Bulk cleaner ingredients should be stored in protected areas, away from heat, water, or whatever could cause the chemical to react. The storage space should have separate areas for acid, alkaline, and reactive cleaners. All storage areas should be fire-safe. Lids must fit tightly, especially if the area has an automatic sprinkler system. All chemicals should be clearly labeled, but especially those that need special warnings.

Workers must completely seal containers of alkaline material between uses, because these chemicals soak up water from the air.

First Aid for Chemical Burns

If cleaning chemicals splash an employee's skin or eyes:

1. *Flush the area with plenty of water immediately. Keep flushing for 15 to 20 minutes.*
2. *Do not use chemicals with the opposite pH to neutralize the burn (i.e., acid chemicals on an alkali burn or alkaline chemicals on an acid burn).* The chemicals may react and have an even more serious effect.
3. Workers can carry a buffered eye solution, sold in sealed containers. If water is not available, the worker can use this solution to dilute and wash away chemicals from their eyes. (They can also carry a plastic squeeze bottle of sterile water.) The employee should then get to a water source as soon as possible and flush his or her eyes for 15 to 20 minutes. Employees should hold their eyes open, and throw handfuls of water into their eyes if necessary.
4. A doctor should always check workers' eyes after they have been in contact with chemicals.
5. Workers should not rely on emergency treatment to protect them. *They should always use the proper equipment to protect their eyes, especially where flushing water is not close by.*
6. *Medical staff should not release an injured employee from first aid or medical treatment until all of the chemical is removed.*
7. It is important to act quickly when an employee has a chemical burn. An employee with a severe burn may be confused and need help. The supervisor or other workers should help the injured employee remove contaminated clothes and flush the area as quickly as possible.

8. Chemical burn showers or eyewash stations are the best places to flush away chemicals. But any source of water, even if it is not clean, is better than not flushing the area. All areas where workmen use corrosive chemicals should have a good water supply. An ordinary showerhead or garden hose spray nozzle does not supply water fast enough to flush away a chemical. A chemical burn shower has a valve that opens as soon as someone steps on a platform or touches another easy-to-use control.

9. After flooding the victim's injury with water, laymen should not attempt other first-aid treatments. A doctor with specialized training in burn treatment should treat all chemical burns. Some chemicals have a toxic effect on the body, and the skin is open to bacterial infections when it is damaged.

10. If the injured person is confused or in shock, keep him still, cover him with a blanket, and take him to a medical facility.

Preventing Skin Problems

Industrial doctors can decide whether employees should be assigned to different tasks because cleaning chemicals irritate their skin. Employees can become sensitive to chemicals even if they have used them for some time without problems. If employees suddenly develop dermatitis, an experienced doctor should examine them. If they have developed a sensitivity, management may need to give the employee different tasks or use a different chemical. Management should keep lists of the chemicals used in the cleaning operation and the recommended treatment for burns both in the first-aid room and in the supervisor's office. The list should include information on how to reach local doctors and medical centers.

CHEMICAL CONTAMINATION OF FOOD

Some of the chemicals (such as cleaning compounds, sanitizers, insecticides, rodenticides, and air fresheners) used in food preparation areas may remain on equipment, utensils, or surfaces, and contaminate the food. Most people have at some time drunk from a glass or cup and tasted the dishwashing soap used to clean it. Insecticides, rodenticides, air fresheners, and deodorizers applied using a spray or vapor may also get into foods. Use of a paint or solid insecticide or pesticide can prevent this problem.

Managers responsible for sanitation can protect equipment and food from chemical contamination by making sure that production and cleanup employees follow procedures. Employees can reduce or even eliminate contamination from food containers and waste if they are careful and vigilant.

SUMMARY

- Sanitation supervisors need to understand the types of soil they need to remove and which chemicals will work best to remove them.

- The best cleaning compound depends on the type of soil. An acid cleaner usually removes inorganic deposits, an alkaline cleaner removes nonpetroleum organic soils, and a solvent-type cleaner removes petroleum soils.
- Cleaning compounds help loosen soil and and flush it away. Cleaning compounds may contain detergent auxiliaries to protect sensitive surfaces or help them clean better.
- Supervisors must know how to handle cleaning compounds to protect employees from injuries.
- If a cleaning compound accidentally splashes onto a worker's skin, eyes, or clothes, the worker should immediately flush the area with plenty of water.

BIBLIOGRAPHY

Anon. 1976. *Plant Sanitation for the Meat Packing Industry.* Office of Continuing Education, University of Guelph and Meat Packers Council of Canada.

Marriott, N. G. 1994. *Principles of Food Sanitation,* 3d ed. Chapman & Hall, New York.

————. 1990. *Meat Sanitation Guide II.* American Association of Meat Processors and Virginia Polytechnic Institute and State University, Blacksburg.

Moody, M. W. 1979. How cleaning compounds do the job. In *Sanitation Notebook for the Seafood Industry,* p. II-68, G. J. Flick et al., eds. Department of Food Science and Technology, Virginia Polytechnic Institute and State University, Blacksburg.

Zottola, E. A. 1973. How cleaning compounds do the job. In *Proceedings of the Conference on Sanitation and Safety,* p. 53. Extension Division, State Technical Services and Department of Food Science and Technology, Virginia Polytechnic Institute and State University, Blacksburg.

STUDY QUESTIONS

1. Give three characteristics of surfaces that are easy to clean.
2. What are the three steps of removing soil?
3. How does mechanical energy (shaking, scrubbing, or high-pressure water) help in cleaning?
4. Why do surfaces have to be clean before they can be sanitized?
5. What is good-quality water for cleaning?
6. Give one example of a type of soil alkali cleaners are good at removing and one example of a type of soil acid cleaners are good at removing.
7. Where and how should cleaning compounds be stored?
8. What is the most important part of first aid for chemical burns?

TO FIND OUT MORE ABOUT CLEANING COMPOUNDS

1. Look at the manuals of various pieces of kitchen equipment (oven, range, refrigerator, mixer, slicer, etc.). Does the manufacturer recommend a specific type of cleaner or cleaning procedure?

2. Read the labels or inserts of cleaning compounds used in the kitchen at work or at home. Is the cleaner acid or alkali? What type of soil does the cleaner remove? Does the cleaner need to be diluted and, if so, how much? Does the manufacturer give any special warnings about the product?

3. Contact your supplier of cleaning chemicals at work (or contact the manufacturer of a cleaner you use at home), and ask if the supplier can send you any information about cleaning compounds.

4. Find and record the telephone number of your local poison control center.

CHAPTER 6

Sanitizing Methods

ABOUT THIS CHAPTER

In this chapter you will learn:

1. The difference between cleaning and sanitizing
2. Why rinsing is important
3. About the three major types of sanitizers: heat, radiation, and chemicals
4. About the most common chemical sanitizers

INTRODUCTION

After food-processing or preparation equipment has been used, it is covered with scraps of food, grease, and other dirt. This dirt usually contains microorganisms and nutrients that allow the microbes to grow. Sanitizing equipment and surfaces requires two steps: cleaning and sanitizing. Cleaning removes the soil deposits, and sanitizing (sterilizing) destroys microbes that are left on the clean surface. If the surface is still dirty, the soil protects the microbes from sanitizing agents. Surfaces must be thoroughly clean for sanitizers to work properly.

HEAT

Heat is an inefficient sanitizer because it takes so much energy. The efficiency of heat depends on the humidity, the temperature required, and the length of time it takes to destroy microbes at that temperature. Heat destroys microorganisms if the temperature is high enough for long enough and if the design of the equipment or plant allows the heat to reach every area. Cleaning staff should use accurate thermometers to measure temperatures during cleaning to make sure that equipment and surfaces are properly sanitized. Steam and hot water are the most common types of heat used for sterilization.

Steam

Sanitizing with steam is expensive because of high energy costs, and it is usually ineffective. Workers often mistake water vapor for steam, so the temperature usually is not high enough to sterilize the equipment or surface. Steam can make bacteria and soil cake onto the surface so that the heat does not reach the microbes; then they stay alive.

Hot Water

Immersing small components (such as knives, small parts, eating utensils, and small containers) into water heated to 82°C (180°F) or higher is another way to sterilize using heat. Pouring hot water into containers is not a reliable way to sterilize, because it is difficult to keep the water hot enough for long enough. Hot water can sanitize food-contact surfaces, plates, and utensils, although spores may survive more than an hour of boiling temperatures.

The time needed to sterilize an item depends on the temperature of the water. If equipment or surfaces are sterilized at a lower temperature, they must be kept at that temperature for longer. If they will be sterilized for a shorter amount of time, the temperature must be higher. This is known as a "time-temperature relationship." Examples of times and temperatures used for sterilization are 15 minutes of heat at 85°C (185°F), or 20 minutes at 82°C (180°F). The volume of water and how fast it is flowing can determine how long it takes for the item being sterilized to reach the right temperature. Hot water is readily available and is not toxic. To sanitize equipment, water can be pumped through it while it is still assembled, or it can be immersed in water.

RADIATION

Radiation in the form of ultraviolet light or high-energy cathode or gamma rays destroys microorganisms. For example, hospitals and homes may use ultraviolet light from low-pressure mercury vapor lamps to destroy microorganisms. Ultraviolet units are already used in Europe and are starting to be used in the United States to disinfect water for drinking and food processing. Generally, the food industry uses radiation to destroy microorganisms in and on fruits, vegetables, and spices; trichina in pork; and *Salmonella* in poultry. It does not work very well in food plants and food-service facilities because the light rays must actually hit the microorganisms and only kill microbes that are very close by. Some bacteria are more resistant to radiation and need a longer exposure for the radiation to destroy them. Dust, grease, and opaque or cloudy solutions absorb radiation and prevent it from killing microbes.

CHEMICALS

Food-processing and foodservice operations use various chemical sanitizers for different areas and types of equipment. More-concentrated sanitizers generally act more quickly and effectively. Sanitation staff need to know and understand how each chemical sanitizer works so that they can choose the best sanitizer for each job. Most chemical sanitizers are liquids, but some chlorine compounds and ozone are gases. It is important not to expose workers to a toxic chemical if a gas sanitizer is used. Chemical sanitizers do not get right into cracks, crevices, pockets, and mineral soils, and so may not completely destroy microbes in these places. When sanitizers are mixed with cleaning compounds, the temperature of the cleaning solution should be 55°C (131°F) or lower, and the soil should be light; otherwise the sanitizer will not work properly. It is also very important to make sure that the chemicals are safe to mix together to avoid dangerous reactions. The effectiveness of chemical sanitizers depends on:

Exposure time. Colonies of microbes die in a logarithmic pattern. This means that if 90% of the microbes die in 10 minutes, 90% of the remaining microbes die in the next 10 minutes, and so on. Therefore, in this example, only 1% of the original number of microbes is still alive after 20 minutes. When more microbes are present, sanitation staff need to use a longer exposure time to reduce the population to a low-enough level. The age of the colony and the type of microorganism affect how quickly they die.

Temperature. Chemical sanitizers kill microorganisms more quickly at hotter temperatures. Bacteria also grow more quickly when the temperature is warmer, but higher temperatures usually speed up their death more than their growth, so that overall microbes die more quickly at hotter temperatures.

Concentration. More-concentrated sanitizers kill microorganisms more quickly.

pH. Even small changes in acidity or alkalinity can affect the activity of sanitizers. Chlorine and iodine compounds are generally less effective when the pH is higher (more alkaline).

Cleanliness. If equipment and surfaces are not thoroughly clean, soil can react with hypochlorites, other chlorine compounds, iodine compounds, and other sanitizers. This reaction neutralizes the sanitizer so that it does not work properly.

Water hardness. Hard water makes sanitizers less effective. The calcium and magnesium salts in hard water neutralize quaternary ammonium compounds. If the water has over 200 parts per million (ppm) of calcium, the sanitation staff should add a sequestering or chelating agent.

Bacterial attachment. Some bacteria attach to solid surfaces. This makes the bacteria more resistant to chlorine.

What to Expect from a Sanitizer

The ideal sanitizer should:

- Destroy all types of vegetative bacteria, yeasts, and molds quickly
- Work well in different environments (i.e., soiled surfaces, hard water, different pHs, soap or detergent residues)
- Dissolve in water

The ideal sanitizer should be:

- Stable as purchased (concentrate) and as used (diluted)
- Easy to use
- Readily available
- Inexpensive
- Easy to measure

The ideal sanitizer should not:

- Irritate skin or be toxic
- Have an offensive odor

One chemical sanitizer cannot sanitize everything. Sanitizers should pass the Chambers test (also called the *sanitizer efficiency test*). This means that the sanitizer should kill 99.999% of 75 million to 125 million *Escherichia coli* and *Staphylococcus aureus* bacteria within 30 seconds at 20°C (68°F). The pH of the sanitizer solution can make it work more or less well. Chemical sanitizers are based on chlorine, iodine, bromine, quaternary ammonium, acid, or other chemicals.

Chlorine Sanitizers

Examples of chlorine-based sanitizers are liquid chlorine, hypochlorites, inorganic or organic chloramines, and chlorine dioxide. These compounds have different antimicrobial activities.

Hypochlorites are the most active of the chlorine compounds and are the most widely used. Calcium hypochlorite and sodium hypochlorite are the most common compounds. Sanitation staff may use calcium hypochlorite, sodium hypochlorite, and chlorinated trisodium phosphate to sanitize after cleaning or add hypochlorites to compatible cleaning solutions for a combined cleaner-sanitizer.

Chlorine dioxide is less effective than chlorine at pH 6.5 (slightly acid), but is more effective at pH 8.5 (alkaline). Therefore, chlorine dioxide may be good for treating sewage. Sanitation personnel can make chlorine dioxide by combining chlorine salt and chlorine (or hypochlorite and acid) and adding chlorite. This sanitizer contains 1 to 5 ppm chlorine dioxide and does not need to be in contact with the surface for as long as quats (quaternary ammonium salts, discussed later) or hypochlorites. Chlorine dioxide works well over the broad range of pHs found in food facilities. It is less corrosive than other chlorine sanitizers because low concentrations work well. This compound also gives off less-unpleasant chlorine smells.

The U.S. Food and Drug Administration (FDA) approved stabilized chlorine dioxide for sanitizing food-processing equipment. Anthium dioxcide is a stabilized solution containing 5% stabilized chlorine dioxide in water with a pH of 8.5 to 9.0. Anthium dioxcide kills bacteria, but it is not nearly as effective as free chlorine dioxide.

Even though chlorine dioxide works at lower concentrations than other sanitizers, it must be generated on-site, and FDA has not yet fully approved its use in food plants. Food processors are using chlorine dioxide more often now that on-site generation equipment is available.

Chlorine concentrations of less than 50 ppm do not destroy *Listeria monocytogenes,* but higher concentrations do. Most chlorine compounds are more lethal to bacteria when more free chlorine is available, the pH is lower, and the temperature is higher. However, chlorine is less soluble in water and more corrosive at higher temperatures, and solutions with a high chlorine concentration and/or low pH (acid) can corrode metals. Chlorine compounds are often preferred over other sanitizers because:

- They include fast-acting compounds that pass the Chambers test at a concentration of 50 ppm.
- They include compounds that kill all types of vegetative cells (i.e., all cells except spores).
- They are usually cheap.
- Employees do not need to rinse equipment if the concentration is 200 ppm or less.
- They are available as liquid or granules.
- Hard water does not usually make them less effective.
- High levels of chlorine can soften gaskets and remove carbon from rubber parts of equipment.

But chlorine-based sanitizers have some disadvantages:

- They are unstable, heat breaks them down, and organic soil makes them less effective.
- Light breaks them down, so they need to be stored in the dark.
- They corrode stainless steel and other metals.
- They can only be in contact with food-handling equipment for a short time, otherwise they corrode dishes or food-handling equipment

It is best to make fresh chlorine solutions just before they are used to make sure they work properly.

Iodine Compounds

The most common iodine-based sanitizers are iodophors, alcohol-iodine solutions, and aqueous iodine solutions. Skin disinfectants often contain these compounds. Iodophors are useful for cleaning and disinfecting equipment and surfaces and in water treatment (see Chap. 8).

Iodine complexes are stable at very low (acidic) pHs, so very low concentrations (6.25 ppm) pass the Chambers test. Iodine sanitizers are more effective than other sanitizers on viruses.

In concentrated form, iodophors have a long shelf life. But once they are dissolved, the iodine may vaporize. Iodine is lost rapidly when the temperature is above 50°C (122°F). Plastic materials and rubber gaskets can absorb iodine and become stained and tainted. Sometimes iodine stain is good because most organic and mineral soil stains yellow, showing up areas that are not thoroughly clean. The amber color of iodine solutions shows when the sanitizer is there, although the intensity of the color is not always related to the concentration.

Iodophor solutions are acidic, so hard water does not make them less effective. Iodophors actually prevent mineral deposits if they are used regularly, although they cannot remove existing mineral deposits. Organic matter (especially milk) absorbs the iodine in iodophor solutions and fades the amber color. The amount of iodine lost from the solution is small unless the organic soil is heavy. The solution loses iodine during storage, so staff should check and adjust the strength of iodine solutions before use.

Iodine compounds have several disadvantages:

- They are more expensive than chlorine compounds.
- They may cause off-flavors in some food products.
- They vaporize at approximately 50°C (122°F).
- They are very sensitive to pH changes.

Iodine is a good hand sanitizer and hand-dipping agent because it does not irritate the skin.

Bromine Compounds

Water treatment facilities use bromine-based sanitizers more often than do food-processing facilities. At a slightly acid or neutral pH, organic chloramine compounds are more effective than organic bromine compounds at destroying spores (such as *Bacillus cereus*). But bromine makes chloramine more effective at alkaline pHs (7.5 or higher). Adding bromine to a chlorine compound solution can make both sanitizers more effective.

Quaternary Ammonium Compounds

The quaternary ammonium compounds, often called the *quats,* are good for cleaning and sanitizing floors, walls, furnishings, and equipment. They are especially good at penetrating porous surfaces. Quats are natural wetting agents and also work as detergents. The most common quats are the cationic detergents, which are poor detergents but excellent germicides. Quaternary ammonium compounds are very effective against *Listeria monocytogenes* and reduce mold growth.

The quats act in a different way from chlorine and iodine compounds against microorganisms. Quats form a bacteriostatic film over surfaces; in other words, they have some residual action. This film is better at killing some bacteria than others. Quats do not kill bacterial spores but can inhibit their growth. Quaternary ammonium compounds work better than chlorine and iodine sanitizers on soiled surfaces, although they work best on surfaces that are thoroughly clean.

Quats are not corrosive, do not irritate the skin, and have no taste or odor when they are properly diluted. The concentration of quat solutions is easy to measure. Cleaning staff should not combine quaternary ammonium compounds with cleaning compounds, because detergent and other ingredients (e.g., anionic wetting agents) may inactivate quats.

Quats are valuable sanitizers because they:

- Are stable and do not react with organic matter
- Do not corrode metals
- Are stable when heated
- Do not irritate skin
- Work at a high pH (alkaline)
- Work well against molds

Disadvantages of quats are that they:

- Do not work so well against certain bacteria
- React with anionic-type synthetic detergents
- Form films on food-handling and food-processing equipment

Acid Sanitizers

Acid sanitizers can combine the rinsing and sanitizing steps. Organic acids such as acetic, peroxyacetic, lactic, propionic, and formic acids are the most common acid sanitizers. The acid neutralizes residues of alkaline cleaning compounds, prevents alkaline deposits from forming, and sanitizes. These compounds work very well on stainless-steel surfaces or where they may be in contact for a long time.

Automated cleaning systems in food plants often combine sanitizing with the final rinse. This has made acid sanitizers popular. Acid sanitizers are sensitive to pH changes, but hard water reduces their effectiveness less than iodine-based sanitizers. Heavy foam used to make it difficult to drain acid sanitizers from equipment, but nonfoaming acid sanitizers are now available. Acids are not as efficient as irradiation, and at high concentrations they can slightly affect the color and odor of foods such as meat. Also, acetic acid does not seem to destroy *Salmonella* bacteria.

Peroxyacetic acid is a rapid sanitizer that kills a wide variety of microbes and is used in food-processing plants. It is less corrosive than iodine and chlorine sanitizers and causes less pitting of equipment surfaces. Peroxyacetic acid is often added to the rinse cycle to reduce the amount of wastewater. It is also biodegradable. Soft-drink manufacturers and brewers use this sanitizer because it works well against yeasts. Peroxyacetic acid is a good sanitizer for aluminum beer kegs. Dairy- and food-processing plants use peroxyacetic acid more and more because it works well against *Listeria* and *Salmonella*.

Acid sanitizers have the following advantages:

- They are not broken down by heat or soil and can be heated to any temperature below 100°C (212°F) without losing power.
- They destroy all vegetative cells (i.e., all cells except spores).
- They are not toxic and are safe to use on most food-handling surfaces.
- They work quickly and kill a wide variety of microbes.

The disadvantages of acid sanitizers are that they are expensive and tend to corrode iron and other metals.

Acid-Quat Sanitizers

Organic acid sanitizers made with quaternary ammonium compounds were marketed as acid-quat sanitizers in the early 1990s. These sanitizers seem to work well, especially against *Listeria monocytogenes*. But they are more expensive than chlorine- or bromine-based sanitizers.

Ozone

Ozone is a possible chlorine substitute. Like chlorine dioxide, ozone is unstable and has to be made as needed on site.

Glutaraldehyde

This sanitizer controls the growth of common bacteria, as well as some yeasts and filamentous fungi found in lubricants used on conveyor belts in the food industry. When sanitation employees add glutaraldehyde to lubricants, bacteria levels fall 99.99%, and fungal levels fall 99.9% in 30 minutes.

Table 6.1 shows the important characteristics of the most common sanitizers. Table 6.2 shows where and how to use sanitizers.

SUMMARY

- Food producers and processors use sanitizers to reduce the number of microorganisms on equipment and surfaces.
- Equipment and surfaces must be thoroughly clean and free of soil for sanitizers to work properly.
- The major types of sanitizers are heat, radiation, and chemicals. Chemicals are more practical than heat and radiation for food production facilities.
- Chlorine compounds are generally the best and cheapest chemical sanitizers, but they irritate skin and corrode surfaces more than iodine sanitizers and quaternary ammonium compounds.
- Bromine compounds are used to treat wastewater more than for sanitizing surfaces, although bromine and chlorine work well together.

- The quats are not such good all-around sanitizers, but work well against molds, have residual activity, and work if the surface is still soiled. They do not kill bacterial spores, but they can slow their growth.
- Food processors may use acid-quat and chlorine dioxide sanitizers more often in the future to kill *Listeria monocytogenes*.

TABLE 6.1. Characteristics of Common Sanitizers

Characteristic	Steam	Chlorine	Iodophors	Quats	Acid
Germicidal efficiency	Good Good	Good	Kills vegetative cells		Selective
Toxic					
Shelf strength	—	Yes	Yes	Yes	Yes
Diluted to use	—	No	Depends on wetting agent	Somewhat	Depends on wetting agent
Stability					
Shelf strength	—	Low	Varies with temperature	Excellent	Excellent
Diluted to use	—	Varies with temperature	Varies with temperature	Excellent	Excellent
Speed	Fast	Fast	Fast	Fast	Fast
Penetration	Poor	Poor	Good	Excellent	Good
Film forming	None	None	None or little	Yes	None
Effect of organic soil	None	High	Moderate	Low	Low
Water properties that affect action	None	Low pH and iron	High pH	Several	High pH
Ease of measurement	Poor	Excellent	Excellent	Excellent	Excellent
Ease of use	Poor	Excellent	Excellent	High foam	High foam
Smell	None	Chlorine	Iodine	None	Some
Taste	None	Chlorine	Iodine	None	None
Effect on skin	Burns	Some	None	None	None
Corrosive	No	Corrodes mild steel	Does'nt corrode stainless steel	No	Corrodes mild steel
Cost	High	Low	Moderate	Moderate	Moderate

Source: Adapted from Lentsch, S. 1979. Sanitizers for an effective cleaning program. In *Sanitation Notebook for the Seafood Industry,* p. 77, G. J. Flick et al., eds. Department of Food Science and Technology, Virginia Polytechnic Institute and State University, Blacksburg.

TABLE 6.2.   Choosing a Sanitizer

Specific Area or Condition	Recommended Sanitizer	Concentration, ppm
Aluminum equipment	Iodophor	25
Bacteriostatic film	Quat	200
	Acid-Quat	Per manufacturer recommendations
	Acid-anionic	100
CIP cleaning	Acid sanitizer	130
	Active chlorine	
	Iodophor	
Concrete floors	Active chlorine	1,000–2,000
	Quat	500–800
Film formation, prevention of	Acid sanitizer	130
	Iodophor	
Fogging, atmosphere	Active chlorine	800–1,000
Hand-dip (production)	Iodophor	25
Hand sanitizer (washroom)	Iodophor	25
	Quat	
Hard water	Acid sanitizer	130
	Iodophor	25
High-iron water	Iodophor	25
Long shelf life	Iodophor	
	Quat	
Low cost	Hypochlorite	
Noncorrosive	Iodophor	
	Quat	
Odor control	Quat	200
Organic matter, stable in presence of	Quat	200
Plastic crates	Iodophor	25
Porous surface	Active chlorine	200
Processing equipment (aluminum)	Quat	200
	Iodophor	25
Processing equipment (stainless steel)	Acid sanitizer	130
	Acid-Quat	Per manufacturer recommendations
	Active chlorine	200
	Iodophor	25

Table 6.2. *(Continued)*

Specific Area or Condition	Recommended Sanitizer	Concentration, ppm
Rubber belts	Iodophor	25
Tile walls	Iodophor	25
Visual control	Iodophor	25
Walls	Active chlorine	200
	Quat	200
	Acid-Quat	Per manufacturer recommendations
Water treatment	Active chlorine	20
Wood crates	Active chlorine	1,000
Conveyor lubricant	Glutaraldehyde	Per manufacturer recommendations

BIBLIOGRAPHY

Anon. 1976. *Plant Sanitation for the Meat Industry.* Office of Continuing Education, University of Guelph and Meat Packers Council of Canada.

Guthrie, R. K. 1988. *Food Sanitation,* 3d ed. Van Nostrand Reinhold, New York.

Haverland, H. 1980. Cleaning and sanitizing operations. In *Current Concepts in Food Protection,* p. 57. U.S. Department of Health, Education and Welfare, Public Health Service, Food and Drug Administration, Cincinnati.

Jowitt, R. 1980. *Hygienic Design and Operation of Food Plant.* AVI Publishing Co., Westport, Conn.

Marriott, N. G. 1994. *Principles of Food Sanitation,* 3d ed. Chapman & Hall, New York.

Marriott, N. G. 1990. *Meat Sanitation Guide II.* American Association of Meat Processors and Virginia Polytechnic Institute and State University, Blacksburg.

Zottola, E. A. 1973. Sanitizers for effective cleaning programs. In *Conference on Sanitation and Food Safety,* p. 59. Extension Division, State Technical Services, Food Science and Technology Department, and Virginia School Food Service Department, Blacksburg.

STUDY QUESTIONS

1. What do you have to do before you can sanitize equipment and surfaces?

2. What is meant by sanitizing?

3. List the three types of sanitizer. Which one is used most often in food-processing facilities?

4. What are the disadvantages of heat sterilization?

5. The food industry uses radiation to destroy microbes in which foods?

6. What are two conditions that must be met if you are going to use a cleaner and sanitizer mixed together?

7. List five things you would look for in an ideal sanitizer.

8. Give two advantages of each of the following sanitizers: chlorine, iodophors, quats, acid.

To Find Out More About Sanitizing Methods

1. Look at the manuals of various pieces of kitchen equipment (oven, range, refrigerator, mixer, slicer, etc.). Does the manufacturer recommend a specific type of sanitizer or sanitizing procedure?

2. Read the labels or inserts of sanitizing compounds used in the kitchen, at work, or at home. Does the sanitizer need to be diluted and, if so, how much? Does the manufacturer give any special warnings about the product?

3. Contact your supplier of sanitizing chemicals at work (or contact the manufacturer of a sanitizer you use at home, e.g., a product containing chlorine), and ask if they can send you any information about sanitizing compounds.

4. Write down each step in the cleaning process for an area or piece of equipment. Mark the steps that clean, those that sanitize, or any that clean and sanitize at the same time.

Helping a Sanitizer Do Its Job

Sanitizers work best when the conditions are just right. This means the right temperature, the right surface, and the right strength or concentration. Researchers in Canada recently looked at how well four sanitizing agents work against *Listeria monocytogenes* with different conditions. The four sanitizers were sodium hypochlorite (chlorine-based), two iodophors, and quaternary ammonium (quat). All of these sanitizers worked better against *L. monocytogenes* on nonporous surfaces than on porous surfaces. The concentration of sanitizers needed to kill *L. monocytogenes* within 10 minutes was 5 to 10 times stronger for a rubber surface (porous) than for a glass or stainless-steel surface (nonporous). Also, sanitizers had to be more concentrated to destroy *L. monocytogenes* at refrigerator temperature (4°C or 39°F) than at room temperature (20°C or 68°F).

Source: Mafu, A. A., Roy, D., Goulet, J., Savoie, L., and Roy, R. 1990. Efficiency of sanitizing agents for destroying *Listeria monocytogenes* on contaminated surfaces. *J. Dairy Sci.* 73:3428.

CHAPTER 7

Cleaning and Sanitizing Systems

ABOUT THIS CHAPTER

In this chapter you will learn:

1. About the costs of different types of cleaning and sanitizing systems
2. How to choose a cleaning system
3. How to monitor a cleaning system to make sure that equipment and surfaces are clean and sanitary
4. About different types of cleaning equipment, including manual equipment, automatic equipment, cleaning-in-place (CIP) systems, and cleaning-out-of-place (COP) systems

INTRODUCTION

Choosing the best cleaning system can be confusing because there are so many types of cleaning equipment, cleaning compounds, and sanitizers to choose from. Truly all-purpose cleaning compounds, sanitizers, or cleaning units do not exist. This is because different types of soil, different surfaces, and different food operations have different needs. This chapter discusses when and how to use different types of cleaning equipment.

Mechanical cleaning and sanitizing equipment is valuable. It can cut down the time spent on cleaning and make cleaning more efficient. A good mechanical cleaning system can cut labor costs by up to 50% and should pay for itself in less than 3 years. A mechanized cleaning unit can also do a better job of removing soil from surfaces than can hand cleaning.

Skilled employees should be chosen to operate cleaning and sanitizing equipment, and qualified managers should supervise these steps. Technical representatives of chemical companies that manufacture cleaning compounds and sanitizers are qualified to recommend cleaning equipment for various uses. But managers should not rely on recommendations from enthusiastic sales representatives. Managers should consider the technology of cleaning and sanitizing problems. They should watch during cleanup to make sure that employees are using cleaning equipment properly.

COSTS

The cost of cleaning is usually divided as follows:

	% of cost
Labor	46.5
Water and sewage	19.0
Energy	8.0
Cleaning compounds and sanitizers	6.0
Corrosion damage	1.5
Miscellaneous	19.0

The largest cost of cleaning is *labor.* About 46.5% of the sanitation dollar is spent on people who carry out and supervise the process of cleaning, sanitizing, and quality assurance. Food processors can reduce this expense by using mechanized cleaning systems.

Water and sewage have the next highest costs. Food plants use a large quantity of water during cleaning and rinsing. Water and sewage treatment are expensive.

It requires *energy* to generate hot water and steam. Energy is expensive. Most cleaning systems, cleaning compounds, and sanitizers work well when the water temperature is below 55°C (131°F). A lower temperature saves energy and makes it less likely that employees will be injured by burns. Lower temperatures also make soil easier to remove, because at high temperatures protein in the soil is denatured (i.e., it changes shape) and sticks more firmly to surfaces.

Cleaning compounds and sanitizers are expensive, but this cost is reasonable because sanitizers destroy microorganisms and allow more thorough cleaning with less labor. The ideal cleaning system uses the best combination of cleaning compounds, sanitizers, and equipment to clean economically and effectively. Using the right amount of cleaning solution and the right concentration helps to control the cost of chemicals.

Corrosion damage from improper use of cleaning compounds and sanitizers on equipment made of stainless steel, galvanized metal, and aluminum costs the industry millions of dollars. Management should buy equipment made out of good construction materials, and employees should use the right cleaning and sanitizing systems to avoid corrosion.

Miscellaneous sanitation costs include equipment depreciation, returned goods, general and administrative expenses, and other operating costs. It is hard to reduce these costs because they are so general. Good overall management can help keep these costs down.

SELECTING A SYSTEM

There are several good sources of information on sanitation systems. These include the planning division (or similar group) of the food company, a consulting organization (internal or external), and the supplier of cleaning and sanitizing compounds and equipment. A basic plan should guide the choice and installation of equipment.

Sanitation Study

A sanitation study should start with a survey of the facility. A study team or individual specialist should note current cleaning procedures (or recommended procedures for a new operation), labor requirements, chemical requirements, and energy and water costs. The team or consultant can then recommend new cleaning procedures, cleaning and sanitizing supplies, and cleaning equipment. Before management changes cleaning systems, the team or consultant should show them that the new system will cost less and/or clean better than the original system.

Using Sanitation Equipment

When management has bought the cleaning equipment, the vendor or another expert should supervise installation and startup of the new system. The vendor or manufacturer should train the staff who will use the system and supervise cleaning. The team or consultant who made the original recommendations should help management inspect the system regularly after startup to make sure it is working well. Supervisors should inspect the system daily, and management should review inspection data every 6 months. Staff should keep records of inspections and reviews (see Chap. 19).

Reports should include how well the program is working, inventory data for cleaning supplies, and the condition of cleaning equipment. Management should compare the costs of labor, cleaning compounds, sanitizers, and maintenance with those projected in the sanitation study. This will help show problems and make sure that actual costs are close to projected costs. Monitored systems can cost up to 50% less than unmonitored systems.

HACCP and Cleaning

HACCP works for cleaning as well as for food production. A sanitation survey can show areas where cleaning is highly critical, critical, or subcritical to prevent physical and microbial contamination of food. Areas can be grouped according to how often they need to be cleaned, as follows:

1. Continuously
2. Every 2 hours (during each break period)

3. Every 4 hours (during lunch break and at the end of the shift)
4. Every 8 hours (end of shift)
5. Daily
6. Weekly

Management should decide on appropriate microbial tests to verify that cleaning is adequate. Staff should collect samples at times and places that will show whether the cleaning system is working. For example:

Flow sheet sampling measures the microbial load of food samples collected after each preparation step. When staff collect samples at each step during production, they can measure the number of microorganisms that come from each piece of equipment.

Environmental samples taken from the food-processing environment are important for controlling pathogens such as *Salmonella* spp. and *Listeria* spp. Staff should collect samples from air intakes, ceilings, walls, floors, drains, air, water, and equipment (including cleaning equipment).

CLEANING EQUIPMENT

Workers usually clean manually, using basic supplies and equipment, or use mechanized equipment, which uses a cleaning medium (usually water), cleaning compound, and sanitizer. Management should make sure the cleaning crew has the tools and equipment they need to clean properly with minimal effort and time. Management should allow storage space for chemicals, tools, and portable equipment.

Mechanical Abrasives

Abrasives such as steel wool and copper chore balls are good for removing soil manually. But cleaning staff should not use these cleaning aids on any surface that has direct contact with food. Small pieces of scouring pads can get stuck in the equipment. They may cause corrosion and pitting (especially on stainless steel), or the food may pick up fragments, leading to consumer complaints and even consumer damage suits. Staff should not use wiping cloths instead of abrasives because clothes spread molds and bacteria. If staff use cloths for any purpose, they should boil and sanitize them before use.

Water Hoses

Hoses should be long enough to reach all areas that need to be cleaned, but should not be longer than necessary. Hoses should have the right nozzles to cover the area for fast and effective cleanup. Nozzles with quick-fit connectors are easiest to use. Fan nozzles cover large surfaces in the least amount of time. Workers can dislodge debris stuck in deep cracks or crevices using small, straight jets. Bent nozzles help to clean around and under equipment. A spray-head brush helps wash and brush at the same time. Cleanup hoses should have an automatic shutoff valve on the operator's end to conserve water, reduce splashing, and make nozzle-changing easier. Workers should remove hoses from food production areas after cleanup. Staff should also clean and sanitize hoses and store them on hooks off the floor. This is especially important for control of *Listeria monocytogenes*.

Brushes

Cleaning brushes should fit the shape of the surface being cleaned. Brushes with spray heads between the bristles can clean screens and other surfaces that need water spray and brushing at the same time. Bristles should be as harsh as possible without damaging the surface. Rotary hydraulic and power-driven brushes for cleaning pipes help clean the inside of lines that transport liquids and heat exchanger tubes.

Brushes are made of a variety of materials: horsehair, hog bristles, fiber, and nylon. Brushes made from bassine, a coarse-textured fiber, are good for heavy-duty scrubbing. Palmetto fiber brushes are less coarse and are good for scrubbing moderately soiled metal equipment and walls. Tampico brushes have fine fibers and are good for cleaning lightly soiled surfaces that only need gentle brushing. Most power-driven brushes have nylon bristles. All nylon brushes have fibers that are strong, flexible, uniform, and durable, and do not absorb water. Brushes made with absorbent bristles are not sanitary.

Scrapers and Squeegees

Sometimes cleaning staff need scrapers to remove stubborn soil. The cleaning employee(s) may use squeegees to clean product storage tanks when the operation is too small to use mechanized cleaning.

High-Pressure Water Pumps

High-pressure water pumps may be portable or stationary, depending on the volume and needs of the food operation. Portable units are usually smaller than centralized units. Portable units pump from 40 to 75 L/min (10–18 gal/min) at pressures up to 41.5 kg/cm^2 (595 lb/in^2). Portable units may have solution tanks that mix cleaning compounds and sanitizers. Stationary units pump from 55 to 475 L/min (15–125 gal/min). Piston-type pumps deliver up to 300 L/min (80 gal/min), and multistage turbines pump up to 475 L/min (125 gal/min) at pressures up to 61.5 kg/cm^2 (882 lb/in^2).

A central high-pressure unit can pipe water throughout the plant, with outlets where staff need to clean equipment or areas. The pipes, fittings, and hoses must be able to withstand the water pressure, and all of the equipment should resist corrosion. Management may choose to use a stationary or portable unit, depending on the volume of high-pressure water that they need and how easily staff can move a portable unit between areas.

Steam Guns

Steam guns mix steam with water and/or cleaning compounds. The best units use plenty of water and are properly adjusted so that steam does not create a fog around the nozzle. This equipment has its value, but it uses a lot of energy and can be unsafe if it causes fog. It causes condensation that can promote growth of mold on walls and ceilings and may allow growth of *L. monocytogenes*. High-pressure water pumps generally work as well as steam guns at a lower water temperature, if staff use the right cleaning compounds.

High-Pressure Steam

High-pressure steam may remove some types of debris and blow water off processing equipment after it has been cleaned. Generally, this is not a good way to clean because it causes fog and condensation and does not sanitize the cleaned area.

Hot-Water Wash

Hot-water washing is a method rather than a type of equipment or a cleaning system. Because only a hose, a nozzle, and hot water are required, this method of cleaning is used a lot. Sugars, certain other carbohydrates, and some salts are relatively soluble in water. Hot water removes these soils better than it removes fats and proteins. Hot-water washing is cheap to install and maintain, but it is not a good cleaning method. Hot water loosens and melts fat deposits, but it denatures proteins so that they change shape and stick to the surface more firmly. Without high pressure, the water does not thoroughly reach all areas, and the method uses a lot of labor unless staff use a cleaning compound. Like other equipment that uses hot water, this method has high energy costs and causes condensation.

Portable High-Pressure, Low-Volume Cleaning Equipment

A portable high-pressure, low-volume unit contains an air- or motor-driven high-pressure pump, a storage container for the cleaning compound, and a high-pressure delivery line and nozzle (Fig. 7.1). The self-contained pump puts pressure into the delivery line; the nozzle regulates the pressure and volume. This portable unit adds the right amount of cleaning compound from the storage container as the pump delivers the desired water pressure. The ideal high-pressure, low-volume unit delivers 8 to 12 l/min (2–3 gal/min) of cleaning solution at about 55°C (131°F) and 20 to 85 kg/cm^2 (285–1,220 lb/in^2) pressure.

The high-pressure spray atomizes the cleaning compound. The velocity, or force, of the cleaning solution against the surface is what makes this an effective cleaning method. High-pressure, low-volume equipment uses less water and cleaning compound. It is also safer than high-pressure, high-volume equipment, because the low volume carries less force, especially farther away from the nozzle.

Portable high-pressure, low-volume equipment is relatively inexpensive and connects to existing utilities. Some suppliers of cleaning compounds provide these units for customers who agree to purchase only their products. These units take more labor than centralized equipment, because staff have to transport them throughout the cleaning operation and because they are less automated than a centralized system. Portable equipment is less durable than centralized equipment and can need a lot of

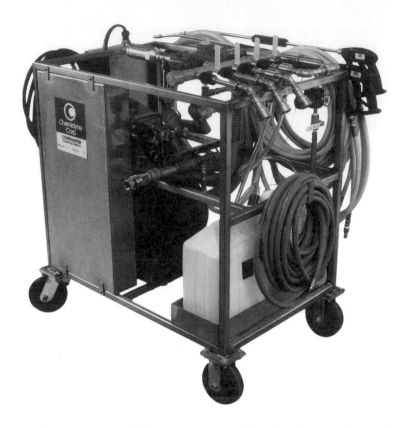

Figure 7.1. A portable high-pressure, low-volume cleaning unit that can be used without a centralized system. This unit has racks for hoses, foaming compounds, and cleaning compounds, and has two rinse stations and a sanitizer unit. Two workers can use the unit at the same time to prerinse, clean, postrinse, and sanitize. This equipment can also apply foam if the spray wand is replaced with a foam wand accessory. *(Courtesy of Chemidyne Corp.)*

maintenance. High-temperature sprays can bake soil onto surfaces and can provide a good temperature for microbes to grow.

But this equipment is useful for small plants because staff can move the portable units throughout the plant. Staff can use portable equipment to clean parts of equipment, building surfaces, conveyors, and processing equipment that cannot be soaked and that would be difficult or time consuming to brush by hand. Managers may use this method more in the future because it seems to help remove *Listeria monocytogenes* from areas that are difficult to clean. However, more and more food processors are likely to use central equipment because it saves labor and needs less maintenance.

Centralized High-Pressure, Low-Volume Systems

This system uses the same principles as portable high-pressure, low-volume equipment. Centralized systems use piston-type or multistage turbine pumps to generate pressure and volume. Like the portable equipment, the high-pressure spray units use the impact energy of the water hitting the soil and surface to clean. The pump(s), hoses, valves, and nozzle parts of ideal centralized high-pressure cleaning systems should be resistant to attack by acid or alkaline cleaning products. Automatic, slow-acting shutoff valves should prevent unwanted spraying and wasting of water. The centralized system is flexible, efficient, safe, and convenient.

But if employees do not use this cleaning system properly, it can blast soil in all directions. Staff should use a low-pressure rinse-down before high-pressure cleaning. Most suppliers of these systems give customers technical assistance and advice about the right cleaning product for the cleaning equipment.

Centralized high-pressure, low-volume systems penetrate and clean similarly to commercial dishwashing machines. The system automatically injects either a detergent disinfectant or a solvent solution into the waterline so that the spray cleans exposed surfaces and reaches into inaccessible or hard-to-reach areas. The spray can flush out cracks and crevices where soil collects and can reduce bacterial contamination. The jet of detergent or detergent-disinfectant solution cuts grease and scours all surfaces. The chemicals clean better because of the water spray. Figure 7.2 shows an example of a high-pressure cleaning system.

Centralized high-pressure cleaning systems work best if all areas that need cleaning have quick connection outlets. The system can dispense several cleaners—acid, alkaline, or neutral cleaners and sanitizers—and belt conveyors can have mechanized spray heads mounted on them with automatic washing, rinsing, and cutoff.

Centralized systems are far more expensive than portable units because they are generally custom built. The cost depends on the size of the facility and how flexible the system is. The initial cost may be from $15,000 to over $200,000.

Choosing Centralized High-Pressure Equipment

The two most common types of centralized equipment are medium pressure (10 kg/cm^2 [145 lb/in^2] with boost to 20 kg/cm^2 [285 lb/in^2]) and high pressure (40 kg/cm^2 [575 lb/in^2] with boost to 55 kg/cm^2 [790 lb/in^2]). Processing plants with heavy soils usually use medium-pressure systems. Beverage and snack food plants that have light soils and need cutting action to clean processing equipment use high-pressure equipment. But managers need to consider several factors before they decide which equipment will give the best long-term results for their operation.

The rinse nozzle flow rate is normally inversely related to the pressure. Each cleaning task requires a certain amount of force to dislodge the soil and flush it off the equipment. At high pressures (40–55 kg/cm^2, 575–790 lb/in^2), the nozzle flow rates are about 5 L/min (1.3 gal/min). But lower pressure needs a high flow rate at the nozzles to achieve the same impact force. Higher flow rates use more water. For example, if a plant has a medium-pressure (20 kg/cm^2, 285 lb/in^2) system with 30 to 40 L/min (7.9–10.5 gal/min) rinse nozzles, and management wants to conserve water, they would need to increase the pressure to 40 to 50 kg/cm^2 (575–715 lb/in^2) and reduce the nozzle flow rate to 10 to 15 L/min (2.6–4.0 gal/min). This would result in the same impact force and use 50% less rinse water. Water conservation also

Figure 7.2. A centralized high-pressure, low-volume system for a large cleaning operation. (*Courtesy of Diversey Corp.*)

means that the plant produces less sewage and uses less energy to heat the water. A high-pressure system may pay for itself in 3 to 6 months.

In the past, plants with a heavy soil have used lower pressure because high pressure tends to splatter soil. But heavier soils need greater impact force. To conserve water, operations should use a pressure of 40 to 50 kg/cm^2 (575–715 lb/in^2) with a nozzle flow rate of 10 to 20 L/min (2.6–5.3 gal/min). If the plant has a very short cleanup period (e.g., only 4 or 5 hours), it may need to use a higher flow rate during this period. Therefore, the system must be able to handle different flow rates.

Centralized high-pressure equipment is a big investment. The cost usually determines what is bought. High-pressure pumps are more expensive than medium-pressure pumps, and all the piping, valves, and other parts cost more because they need to withstand the higher pressure. Usually, savings from water conservation cover setup and operating costs over time.

Medium-pressure equipment is less expensive to buy and use. The pumps are less sophisticated, and all the piping, valves, and other parts do not need to withstand such high pressure. Maintenance costs are also lower than for high-pressure systems.

If water use is not a big issue for the operation, a medium-pressure system is probably best. Many processors use a pressure of 20 kg/cm^2 (285 lb/in^2) with 20 to 30 L/min (5.3–7.9 gal/min) nozzles in most areas of the plant. Supervisors must train employees to use the equipment properly and understand sanitation procedures.

Portable Foam Cleaning

Foam is easy and quick to use, so foam cleaning has been popular during the past decade. The cleaning compound mixes with water and air to form the foam. Clinging foam is easy to see, so the worker knows where he or she has used the cleaning compound and is not likely to clean the same area twice.

Foam cleaning works especially well for large surface areas because it clings to the surface, so that the cleaning compound has a longer time to act. Workers often use foam to clean the inside and outside of transportation equipment, ceilings, walls, piping, belts, and storage containers. The equipment is about the same size and costs about the same as portable high-pressure units. Figure 7.3 shows a portable foam cleaning unit.

Centralized Foam Cleaning

This equipment uses the same technique as portable foam equipment, except that the plant has several carefully planned stations for quick connection to a central foam gun. Centralized equipment has similar advantages to the centralized high-pressure system. The food industry now uses more centralized rather than portable foam-cleaning equipment because the centralized system saves labor and needs less maintenance. As with portable foam cleaning, the cleaning compound mixes with water and air to form a foam, which is piped to various stations throughout the plant. Figures 7.4, 7.5, and 7.6 show various types of centralized foam equipment.

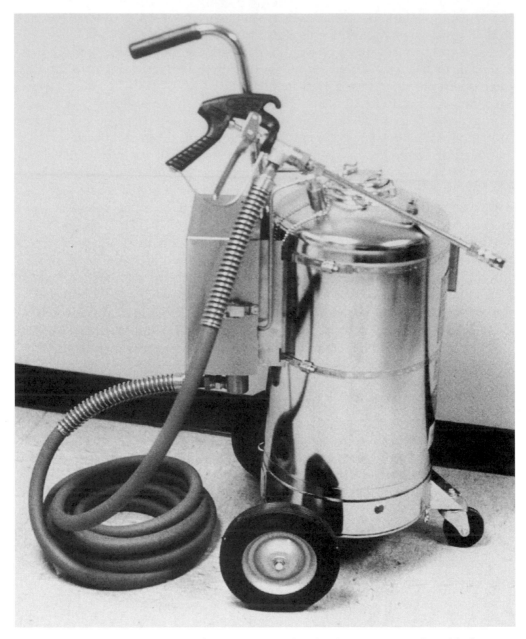

FIGURE 7.3. A portable air-operated foam unit that applies the cleaning compound as a blanket of foam. *(Courtesy of Diversey Corp.)*

The compact wall-mounted foam generation unit shown in Figure 7.5 is designed to mix cleaning compounds straight from the original shipping containers. Planners can mount wall units in areas where staff do a lot of cleaning. The equipment shown in Figure 7.5 can blend and dispense cleaning compounds through an

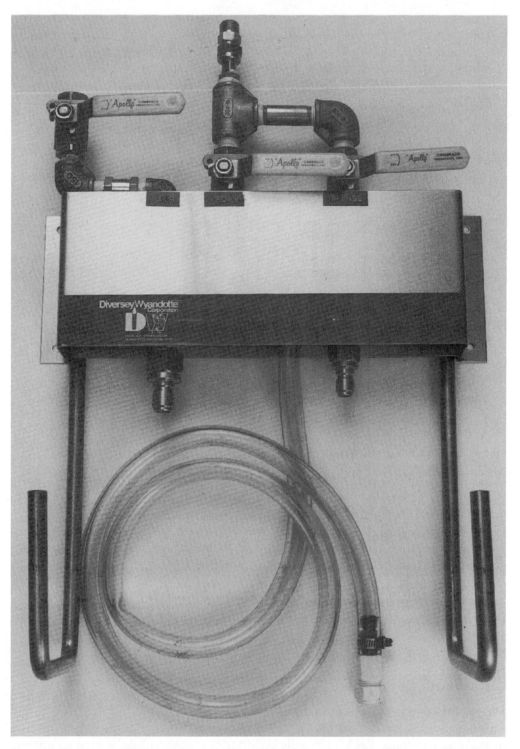

FIGURE 7.4. A centralized cleaning unit that contains two rinses, two combination foamer-spray washers, and a sanitizer. (*Courtesy of Oakite Products, Inc.*)

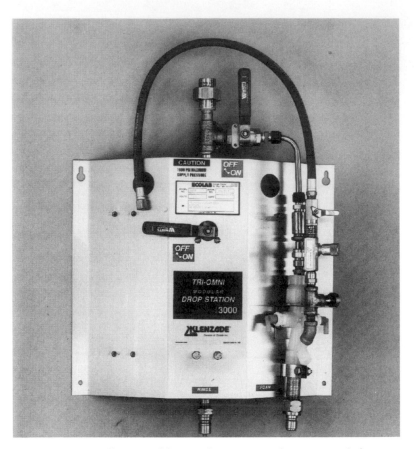

FIGURE 7.5. A wall-mounted foam and rinse station that can provide foam application at convenient locations in a food plant through automatic metering and mixing of the cleaning compound. *(Courtesy of Klenzade—a service of Ecolab.)*

adjustable air regulator and water-metering valve. The controls that monitor the amount of chemicals, air, and water are in the latching stainless-steel cabinet and are easy to get to. This equipment has a built-in vacuum breaker and check valves in the air and water lines.

Figure 7.6 shows drop stations that can dispense foam, high-pressure water for rinsing, and a sanitizer. The foam station has adjustable air and detergent regulators to create the proper foam. The rinse unit can provide up to 69 kg/cm^2 (990 lb/in^2) of pressure.

Portable Gel Cleaning

This system is similar to portable high-pressure units, except that no air is mixed with the cleaning compound, and it is used as a gel rather than a foam, high-pressure spray, or other media. Gel is especially good for cleaning food-packaging equipment because the gel clings to the moving parts and helps remove the soil. The equipment costs about the same as portable foam and high-pressure units.

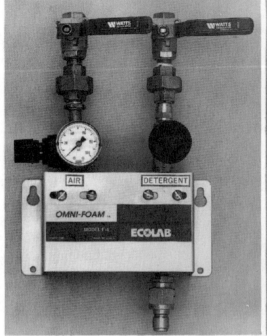

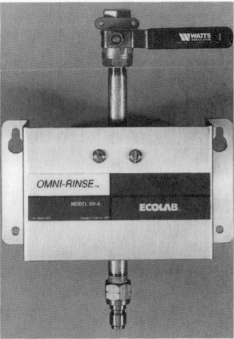

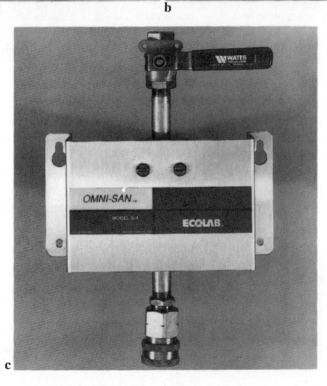

FIGURE 7.6. These drop stations provide a quick connection for the (a) foam detergent, (b) rinse, and (c) sanitizer applications. *(Courtesy of Klenzade—a service of Ecolab.)*

Centralized or Portable Slurry Cleaning

This method is the same as foam cleaning, except that it mixes less air with the cleaning compounds. It forms a slurry that is more fluid than foam and penetrates uneven surfaces better. Compounds applied as a slurry are not in contact with the surface for as long as foam, because the foam clings to surfaces better.

Combined Centralized High-Pressure and Foam Cleaning

This system is the same as centralized high-pressure cleaning, except that it can use high pressure and foam. This system is more flexible than most cleaning equipment because staff can use the foam to clean large surface areas and apply high pressure for belts, stainless-steel conveyors, and hard-to-reach areas. These systems are expensive (from $15,000 to over $200,000, depending on the size) because they usually have to be custom designed and built.

Cleaning in Place (CIP)

Labor continues to cost more, and hygienic standards are becoming stricter. Therefore, CIP systems are more and more valuable. Dairies and breweries have used CIP for many years, but few other plants have used it because it is expensive to buy and install, and some processing equipment can be hard to clean this way. Therefore, food processors can only use CIP in certain places, and the system must be custom designed. CIP equipment is best for cleaning pipelines, vats, heat exchangers, centrifuges, and homogenizers used to process liquids.

Custom-designed CIP equipment may be more or less automated with anything from simple timers to fully automated computer-controlled systems. The choice depends on the equipment budget, labor costs, and the type of soil. A reliable consulting firm or a reputable equipment and detergent supplier should design the system. These organizations can survey the site and provide confidential reports on the hygienic status of the current equipment and cleaning techniques.

Small-volume plants cannot always justify full automation. If the system has less automation, the circuits can be set manually. If the system has full automation, the entire CIP operation can be automatic. Safeguards prevent dangerous valve errors.

The CIP principle combines the benefits of the chemical activity of the cleaning compounds with the physical effects of sprays, flowing fluids, and brushes. The system dispenses the cleaning solution onto the soiled surface, for the right amount of time, at the right temperature, at the right concentration, and with the right amount of force. For this system to work, a relatively high volume of solution has to be in contact with the surface for at least 5 minutes and sometimes as long as 1 hour. The system recirculates the cleaning solution to conserve water, energy, and cleaning compounds.

CIP systems now permit water from the final rinse to make up the solution for the next cleaning cycle to use less freshwater and produce less wastewater. The dairy industry has tried various other ways to reuse water and chemicals and use less energy by filtering, evaporating, and recycling.

Properly designed CIP systems can clean some equipment in food plants just as well as when it is dismantled and cleaned by hand. In many food plants, CIP equipment has completely or partially replaced hand cleaning.

The simplified flowchart in Figure 7.7 shows how a CIP system works. The arrangement shows mixing and detergent tank(s), pipelines, heat exchanger(s), and storage tank(s). A CIP system can clean storage tanks, vats, and other storage containers. It can also clean pipelines and various plant items using a high-velocity cleaning solution of water and cleaning compounds. The solution is recirculated. A typical cleaning cycle for a CIP system is listed in Table 7.1.

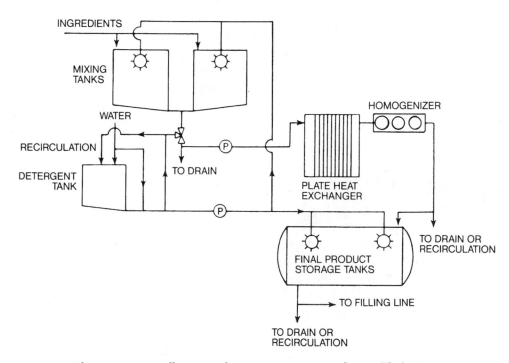

FIGURE 7.7. Flow arrangement illustrating the operation sequence of a simplified CIP system. *(From Jowitt, 1980.)*

TABLE 7.1. Typical Cycle for a CIP System

Step	Reason
1. Preliminary rinse (hot or cold water)	Remove most soil
2. Detergent wash	Remove remaining soil
3. Rinse	Remove cleaning compounds
4. Sanitize	Destroy microorganisms
5. Final rinse (optional: depends on sanitizer used)	Remove CIP solutions and sanitizers

Plant layout is an important part of CIP systems because dismantling of equipment is unnecessary. Ideally, the pipework and joints have crevice-free joints and smooth walls and can be cleaned by liquid spray. Spray balls are either fixed or rotate and spray a high-speed jet of liquid over the entire interior of the tank to thoroughly remove soil, other contaminants, and microbes. Figure 7.8 shows a spray ball.

The two basic CIP designs are single-use and reuse systems. Multiuse systems combine the best characteristics of the single-use and reuse equipment.

Single-Use Systems

Single-use systems use the cleaning solution only once. Single-use systems are usually small and are usually put right next to the equipment they clean and sanitize. Because the units are close to the equipment they clean, they can use a relatively small amount of chemicals and rinse water. Single-use systems work well for heavily soiled equipment because solutions cannot be reused when the soil is heavy. Some single-use systems recover the cleaning solution and rinse water and use it to pre-rinse in the next cleaning cycle.

FIGURE 7.8. A spray ball that mechanically cleans large and small tanks. It consists of a stainless-steel head on a self-cleaning alkali- and acid-resistant ball race. The spherical head has a wide-angle nozzle and is slotted to saturate surfaces in every direction. An offset jet spins the ball vertically at adjustable speed. (*Courtesy of Oakite Products, Inc.*)

Single-use units are more compact and less expensive to buy than other CIP systems. These units are also less complex and easier to install. Figure 7.9 shows a typical single-use system.

A typical cleaning sequence for equipment such as storage tanks or other storage containers takes about 20 minutes. This includes the following steps:

1. Three prerinses of 20 seconds each with intervals of 40 seconds between them remove the gross soil deposits. The water is subsequently pumped to the drain by a CIP return pump.
2. The cleaning solution is injected with steam (if used) to heat it to the right temperature on its way into the circuit. The cleaning solution washes the equipment for 10 to 12 minutes, and then the spent chemicals go to the drain or recovery tank.

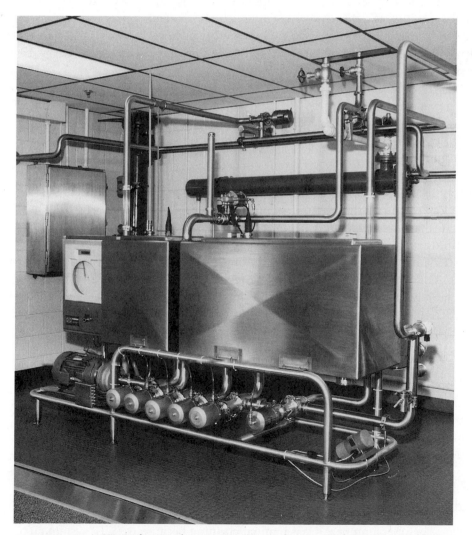

FIGURE 7.9. A CIP single-use solution recovery unit that is part of a system containing a water supply tank and CIP circulating unit. *(Courtesy of Klenzade—a service of Ecolab.)*

3. Two intermediate rinses with cold water for 40 seconds each remove the cleaning chemicals. The rinse water goes to the recovery tank or drain.

4. A final rinse may include acid to lower the pH to 4.5. The rinse water circulates cold for about 3 minutes and then goes to the drain.

Reuse Systems

Reuse CIP systems are important to the food industry because they recover and reuse cleaning compounds and the cleaning solution. It is important to understand that cleaning solutions are hardly contaminated because the prerinse removes most of the soil, so that the system can use the cleaning solutions more than once. For this system to work, cleaning staff must use the proper concentration of the cleaning solution. The chemical supplier and the equipment vendor have guidelines to make sure that the concentration is right.

Reuse CIP systems have a tank for each chemical. A hot-water tank or a bypass loop helps to save energy and water during a hot-water rinse. A coil often heats the cleaning solution.

The basic parts of a CIP reuse system are an acid tank, alkali tank, freshwater tank, return-water tank, heating system, and CIP feed and return pumps. Remote-controlled valves and measuring devices come with the piping layout of this cleaning system. The cleaning steps are automatic, and cleaning staff set them using a program control unit.

Two-tank systems use one tank to reclaim rinse water and another to reclaim the cleaning solution. CIP equipment with three tanks has one tank for the cleaning solution, one for reclaiming the prerinse solution, and one for a freshwater final rinse. Both single-use and reuse systems need good design and monitoring to make sure that food products never mix with cleaning solutions.

CIP systems may have two tanks for solutions of alkaline cleaning compounds at two different concentrations. The less-concentrated solution cleans tanks, other storage parts, and pipelines. The stronger solution cleans the plate heat exchanger. Pumps that feed the cleaning compounds into the tanks automatically neutralize the alkaline pH with the right amount of acid.

A CIP system can clean two circuits at the same time by adding extra feed pumps. The size of the tanks depends on the volume of the circuit, the temperature that is needed, and the cleaning program. In mechanized plants, a central control console contains remote-controlled valves that switch the cleaning circuits on and off. A return water tank can cut down water consumption in a reuse system. Reuse equipment costs more initially but saves money over time.

The ideal CIP reuse system fills, empties, recirculates, heats, and dispenses automatically. Table 7.2 shows how a typical system works to clean storage tanks and pipelines and recover the cleaning solution.

The advantages of CIP equipment are:

- *Less labor.* The operation needs less manual cleaning because the CIP system automatically cleans equipment and storage utensils. Higher wages and more difficulty finding good workers have made labor saving a greater priority.

TABLE 7.2. Steps in an Ideal CIP Reuse System

Step	Time, min	Temperature
Prerinse: Cold water comes from the recovery tank and goes to the drain.	5	Room temperature
Detergent wash: A 1% alkaline base cleaning compound sends the rest of the rinse water to the drain, and then a probe diverts it to the cleaning compound tank for recovery and recirculation.	10	Room temperature to 85°C (185°F) depending on the equipment and type of soil
Intermediate water rinse: Softened cold water from the rinse forces the rest of the cleaning solution out to the cleaning solution tank. The rest of this water goes to the water recovery tank.	3	Room temperature
Acid wash: A 0.5–1.0% acid solution forces out the rest of the water to the drain, and then a probe diverts it to the acid tank for recovery and recirculation.	10	Room temperature to 85°C (185°F) depending on the equipment and type of soil
Final water rinse: Cold water washes out the rest of the acid solution, and water is collected in the water recovery tank. Overflow goes to the drain.	3	Room temperature
Note: Pasteurizing equipment, tanks, and pipelines may also have a final rinse of hot water at 85°C (185°F).		

Source: Jowitt (1980).

- *Better hygiene.* Automated systems clean and sanitize more effectively and consistently than people do. Timers and computer-driven controls precisely control cleaning and sanitizing operations.
- *Less cleaning solution.* Automatic meters and reuse of water, cleaning compounds, and sanitizers avoid waste and save money.
- *Better use of equipment.* Automated systems can clean equipment, tanks, and pipelines as soon as production finishes, so operators can reuse the equipment right away.
- *Better safety.* Workers do not need to get into equipment to clean it, so they are less likely to slip on internal surfaces.

The disadvantages of CIP systems are:

- *Cost.* Most CIP systems are expensive because they are custom designed and expensive to install.
- *Maintenance.* More sophisticated equipment and systems need more maintenance.

- *Inflexibility.* These systems only clean the areas where the equipment is installed. Portable cleaning equipment can cover more areas. CIP systems do not clean heavily soiled equipment well, and it is difficult to design units that can clean all types of processing equipment.

Microprocessor Control Unit

Sophisticated CIP equipment now includes a microprocessor control unit. Workers can program the unit to operate different single-use or reuse cycles, so the system is more flexible. The CIP unit produces a graphic record showing temperatures, pressures, flows, pHs, concentrations, valve openings and closings, and other data. These data help operators make sure the system is working properly and show when it needs maintenance.

The microprocessor control unit makes the cleaning unit work better and helps keep cleaning costs down by controlling the use of energy, water, and chemicals. One unit can have up to 200 programs that allow for the system to reuse water or cleaning compounds, rinse manually, add a sanitizing cycle, vary the chemical concentrations, extend the wash time, and many other options. Workers can often program the microprocessor control unit using a key pad attached to the system or using a computer connected to the system.

Cleaning Out of Place (COP)

When using a COP system, workers must take equipment apart or move it from the production area to a cleaning area. In the past, regulatory agencies have used fluid flow force to judge whether a cleaning system is adequate. A fluid flow speed of 1.5 meters per second (5 ft/sec) was considered adequate. But COP equipment can clean properly with lower fluid flow speeds, so this standard is no longer necessary.

Cleaning staff can wash many small parts of equipment, utensils, and small containers in a recirculating-parts washer, also called a COP unit. These units have a pump that recirculates the cleaning solution and rinse water, and distribution headers that agitate the cleaning solution. Operators can also use a COP unit to recirculate solutions for a CIP system. The normal wash cycle takes about 30 to 40 minutes, and the cold acid or sanitizing rinse takes about 5 to 10 minutes more.

A COP unit often has a double-compartment stainless-steel sink with motor-driven brushes. The same motor pumps a cleaning solution through a perforated pipe onto the brushes. A heater with a thermostat controls the temperature of the cleaning solution (45–55°C, 113–131°F). The first compartment of the sink contains the cleaning solution. The second compartment contains the rinse solution and a spray nozzle. The clean equipment and utensils air-dry within the COP unit or on a drainer or rack.

COP equipment often contains a brush assembly, a rinse assembly, a tank to hold the cleaning solution, and rotary brushes for cleaning the inside and outside of parts and utensils. The cleaning solution comes out of the pipes connected to the brushes that clean the insides.

The big benefit of a COP system is that it can clean so many different things: parts of equipment, small equipment, and utensils. This system also uses less labor and is more hygienic than washing by hand. COP units are fairly inexpensive to buy and maintain. However, small-volume operations may still find that the costs of buying and maintaining the equipment are high and that they take a lot of labor to load and unload.

The food preparation and foodservice industries often use COP systems to clean equipment and utensils. Chapters 11 and 17 give more information on COP in the dairy and foodservice industries.

SANITIZING EQUIPMENT

Sanitizing equipment includes hand sprayers (e.g., to apply insecticides and herbicides), wall-mounted units, and spray heads mounted on processing equipment. Many mechanized cleaning units include sanitizing equipment as part of the system (see Fig. 7.10).

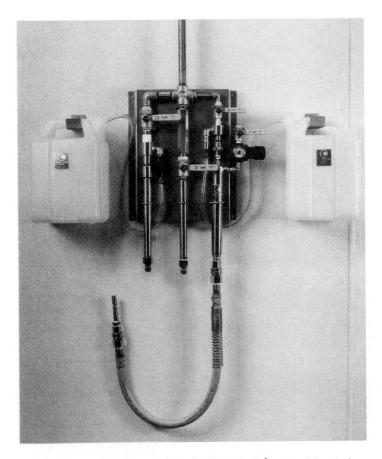

FIGURE 7.10. Wall-mounted combination rinse-foam-sanitize station. *(Courtesy of Chemidyne Corp.)*

Centralized high-pressure, low-volume cleaning and foam cleaning equipment include sanitizing lines. Workers use hoses and wands to apply the sanitizer or headers to spray the sanitizer onto processing equipment such as moving belts or conveyors. Mechanized sanitizing is good because it is uniform, timers can control it, and a meter can control the amount of sanitizing compounds.

Figure 7.11 illustrates a wall-mounted sanitizing unit. The unit controls the flow rate of high-pressure rinse water. An injector meters a specific amount of sanitizer into the high-pressure water stream. A flood nozzle can spread the sanitizing solution without atomizing it.

LUBRICATION EQUIPMENT

Figure 7.12 shows typical equipment for continuous lubrication of conveyors or chains. This equipment may lubricate high-speed bottling and canning conveyors in the beverage industry, shackle chains and conveyors, smokehouse chain drives, and other similar equipment. Water under pressure drives a piston that draws lubricant from a drum and injects it into the water.

SUMMARY

- Cleaning equipment uses cleaning compounds, sanitizers, water pressure, and brushes to help clean and sanitize equipment, utensils, and surfaces.
- A good cleaning system cuts labor costs up to 50%.

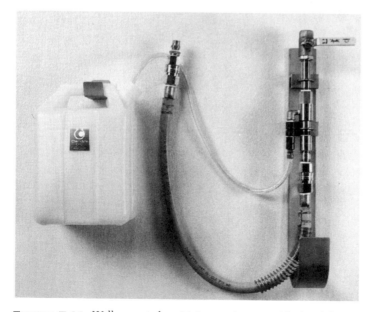

FIGURE 7.11. Wall-mounted sanitizing equipment with the ability to flood sanitize a large area. *(Courtesy of Chemidyne Corp.)*

FIGURE 7.12. Lubrication equipment for high-speed conveyors, drives, and shackle chains. *(Courtesy of Chemidyne Corp.)*

- High-pressure, low-volume cleaning equipment is usually best for removing soil deposits from hard-to-reach areas.
- Foam has become more popular because it is easy to use, clings to surfaces, and works well for large surface areas. Slurries are similar to foam, except that they contain less air and do not cling so well. A gel is best for cleaning packaging equipment.
- Labor-saving CIP units work well for cleaning some equipment used in food-processing plants. CIP systems work well for fluid-processing equipment, e.g., in the dairy and beverage industries. However, this equipment is expensive and does not work well on heavy soil. Sophisticated CIP equipment uses a microprocessor control unit to control the system.
- COP equipment works well for small parts and utensils.
- Mechanized lubrication equipment allows sanitary lubrication of high-speed conveyors and other equipment.

BIBLIOGRAPHY

Anon. 1976. *Plant Sanitation for the Meat Packing Industry.* Office of Continuing Education, University of Guelph and Meat Packers Council of Canada.

Guthrie, R. K. 1988. *Food Sanitation,* 3d ed. Van Nostrand Reinhold, New York.

Jowitt, R. 1980. *Hygienic Design and Operation of Food Plant.* AVI Publishing Co. Westport, Conn.

Marriott, N. G. 1994. *Principles of Food Sanitation,* 3d ed. Chapman & Hall, New York.

Marriott, N. G. 1990. *Meat Sanitation Guide II.* American Association of Meat Processors and Virginia Polytechnic Institute and State University, Blacksburg.

STUDY QUESTIONS

1. Almost half of the money spent on cleaning is spent on _____.
2. Suggest three ways that operators can lower the cost of cleaning.
3. Why should staff not use manual abrasives on food equipment?
4. What are the disadvantages of cleaning and sanitizing with steam?
5. What are two benefits of high-pressure equipment?
6. What are the advantages and disadvantages of centralized and portable cleaning equipment?
7. Why is foam cleaning popular?
8. Compare CIP and COP.

TO FIND OUT MORE ABOUT CLEANING AND SANITIZING SYSTEMS

1. Read the manuals of various pieces of cleaning equipment at work and at home (e.g., COP, CIP, portable or centralized cleaning units, dishwasher, foam carpet cleaner).
2. Talk to managers in charge of sanitation at your workplace or at a local food-processing plant or foodservice establishment. Ask what types of cleaning systems they use and why.
3. Contact your supplier of cleaning chemicals at work (or contact the manufacturer of a cleaning compound you use at home), and ask if the supplier has information about which cleaning compounds work best with different types of cleaning equipment.
4. Write down each step in the cleaning process for an area or piece of equipment. What type of equipment do cleaning staff use for each step? Could the facility use a better system, based on the information in this chapter?

WHY ARE CLEANING CLOTHS HAZARDOUS?

Cleaning cloths carry microorganisms. When workers use cloths in different areas of the kitchen or on different pieces of equipment, they can spread microbes from one area to another or from workers hands to utensils. Wet cloths carry more microorganisms than dry cloths, but even on dry cloths microorganisms can survive for several hours. One study showed that most bacteria live for up to 4 hours on dry cloths, and some survive for up to 24 hours. Contaminated cloths transferred enough microorganisms to fingers, surfaces, or equipment to cause an outbreak of foodborne disease in some situations. Workers should change cleaning cloths often and should use a separate cloth for each piece of equipment or area. Cleaning staff should wash and sterilize cleaning cloths. Equipment, utensils, and surfaces should air-dry.

Source: Scott, E. and Bloomfield, S. F. 1990. The survival and transfer of microbial contamination via cloths, hands and utensils. *J. Applied Bacteriol.* 68:271

CHAPTER 8

Waste Disposal

ABOUT THIS CHAPTER

In this chapter you will learn:

1. Why food production facilities need to be concerned about waste disposal
2. How a facility can decide which waste-disposal systems will work best and be the least expensive
3. How to test the level of pollution in wastewater
4. How food processors handle solid waste
5. How food processors and municipal water treatment systems handle wastewater using (1) pretreatment; (2) primary, secondary, and tertiary treatments; and (3) disinfection

INTRODUCTION

Waste materials from food-processing and foodservice facilities can be hard to treat because they contain large amounts of carbohydrates, proteins, fats, and minerals. Untreated wastes from dairy plants, food-freezing and dehydration plants, and processing plants for red meats, poultry, and seafood can smell unpleasant and pollute water. Processors need to treat and biologically stabilize the organic matter in these wastes before they discharge it into rivers, lakes, or oceans. Improper waste disposal is hazardous to humans and aquatic life.

Federal, state, and local regulatory agencies and the public are demanding better treatment of industrial waste. Processors and regulatory agencies are responsible for prompt and complete disposal of waste materials. When waste accumulates, even for short periods of time, it attracts insects and rodents, smells unpleasant, becomes a public nuisance, and looks unattractive.

The organic matter in waste is an excellent food source for microbes. With such a good food supply, microorganisms grow rapidly and use up the oxygen in the water. Water normally contains about 8 parts per million (ppm) of dissolved oxygen. Fish need at least 5 ppm of dissolved oxygen to live. If the amount of oxygen falls below this level, the fish suffocate. When this happens, the water becomes septic, smells foul, and looks dark. When the waste contains sulfur-containing proteins or a high sulfate content, the septic water releases hydrogen sulfide. Hydrogen sulfide smells like rotten eggs and can blacken buildings.

The content of organic matter in waste from food-processing and foodservice operations is measured as biochemical oxygen demand (BOD). BOD is the amount of oxygen the waste uses as it decomposes. Facilities that put waste with a high BOD into a municipal treatment system have to pay a surcharge because the wastewater needs more treatment. Many large-volume firms treat their own waste either partially or completely to reduce the burden on local water treatment facilities.

STRATEGY FOR WASTE DISPOSAL

Food operators need to conduct a waste-disposal survey to find out how much and what type of waste the facility produces.

Planning the Survey

The first step in a waste-disposal survey is finding the sources of wastes. Construction drawings of the plot plan, piping plans, and equipment layouts can show where water goes in and out. The piping plans should show waterlines, storm sewer lines, sanitary sewer lines, and processing waste drains and lines. The drawings should also show the pipe sizes, pipe locations, types of connections to processing equipment, and flow direction.

The survey should include the operating schedule of the food plant, i.e., the number of shifts and volume of products produced in a day, a week, a month, a season, and the entire year. Managers conducting the survey should look at production records for several years. They should also look at water use records.

An initial waste survey shows whether a plant meets federal, state, and local effluent requirements. This is needed to get or keep a National Pollution Discharge Elimination System (NPDES) permit. The NPDES permit gives responsibility for monitoring the waste stream to the firm that produces it. The Environmental Protection Agency (EPA) checks waste output every so often to check that the firm's reports are accurate.

An initial survey also shows the kind of monitoring equipment the system will need and where to place it so that the operator can monitor waste output continuously. The survey shows whether the operator needs to treat wastes to meet regulations and the best type of waste treatment system.

Conducting the Survey

Information from the operations study shows what to include in the survey. The operator may have to conduct surveys in each season if the plant processes different types and amounts of products at different times of the year. This is common, especially in fruit- and vegetable-processing plants. The survey must include these steps: checking water balance, sampling wastewater, and checking the amount of pollution.

Checking water balance. Operators must place meters on all incoming waterlines to measure the amount and flow speed of wastewater. Examples of measuring devices are Parshall flumes, rectangular and triangular weirs, and venturi tubes and orifices. Operators calculate water balance for an entire plant to give them information on the amount of water in the waste effluent; the amount lost through steam leaks, evaporation, and other losses; and the amount used to make food products in the plant. When operators add these amounts together, they should add up to the amount of water supplied to the plant. This calculation can identify problems such as hidden water losses or major leaks. These problems can stop the sanitation program from working properly and cause more waste, more effluent discharge, and lower profits.

Sampling wastewater. The size of waste samples should be proportional to the rate of flow of the waste. Random samples have little value for checking the amount and type of wastes, and results from these samples can be misleading. For valid data, operators should use a statistical sampling plan to time sample collections while the facility is operating and while it is idle.

The sampling equipment should be in the wastewater system to collect a representative sample (one that gives a true picture of the wastewater at that time). Operators should collect samples at places where wastes are homogeneous (i.e., they are mixed up enough that a sample will show what the entire waste stream is like). Good places may be below a weir or flume. Operators should be careful to avoid sampling errors due to solid deposits upstream from a weir or from grease buildup immediately downstream. They should collect the sample near the center of the channel and at 20 to 30% of the depth below the surface, where the water moves quickly enough that solids are not deposited. Sewers and deep, narrow channels should be sampled at 33% of the water depth from the bottom to the surface. Operators should not agitate samples, because this releases dissolved oxygen from the water. Food plant wastes break down quickly at room temperature, so staff should chill samples to 0 to 5°C (32–41°F) if they do not analyze them immediately.

Checking the amount of pollution. Much of the waste from fruit and vegetable plants, animal slaughterhouses, and cleanup water discharge is pieces of food products. Screens can remove larger pieces. Finer solids pass through the screen, and the organic matter in the wastewater usually has a higher oxygen demand than the amount of oxygen dissolved in the water.

Biochemical oxygen demand. The 5-day BOD test (BOD_5) measures pollution strength. The BOD of sewage and industrial wastewater is the oxygen (in parts per million) required to stabilize the organic matter while aerobic microorganisms decompose it. The technician stores the sample in an airtight container for a certain length of time at a certain temperature. At 20°C (68°F) the sample may take more than 100 days to stabilize completely.

Domestic sewage with no industrial waste has a BOD of about 200 ppm. Food-processing wastes normally have a higher BOD, often more than 1,000 ppm. Table 8.1 shows the typical BOD_5 values and the amount of suspended solids in wastes from various industries. BOD_5 values are usually related to levels of suspended solids, but not to levels of dissolved solids.

TABLE 8.1.   Typical Composition of Wastes from Food and Related Industries

	Parts per million	
Type of Waste	BOD_5	Suspended Solids
Dairy and milk products	670	390
Food products	790	500
Glue and gelatin	430	300
Meat products	1,140	820
Packinghouse and stockyards	590	600
Rendering	1,180	630
Vegetable oils	530	475

Although operators often measure water pollution using BOD_5 and the test is easy to do, it is time consuming and does not give consistent results. Tests such as chemical oxygen demand (COD) and total organic carbon (TOC) are quicker, and the results are more reliable and consistent.

Chemical oxygen demand. The COD test measures pollution chemically rather than biologically. BOD only measures degradable materials, but COD also measures nondegradable materials because it is a chemical test. When wastewater from a food-processing plant is treated by the municipal water treatment system, daily COD measurements can show if and when the wastewater could create a problem at the wastewater treatment plant. However, this test does not show whether microorganisms can degrade the organic matter biologically and, if so, how quickly. Results of this test are similar to but not the same as results from the BOD_5. Results of the COD test are related to the level of dissolved organic solids. In the past, regulatory agencies have not accepted COD data as a substitute for BOD data, unless the plant has recorded a consistent ratio between COD and BOD values.

Dissolved oxygen. Operators need to know the dissolved oxygen (DO) content of both incoming water and wastewater, because it affects aquatic life and is important in water treatment systems. Electrode probes can show DO concentrations.

Total organic carbon. Total organic carbon (TOC) measures all organic materials. Solid matter from the wastewater undergoes catalytic oxidation at 900°C, and the test measures the amount of carbon dioxide that is released. This pollution measurement is fast and consistent, and gives results similar to BOD_5 and COD tests. However, the test is difficult to do and needs special laboratory equipment. This test works well if the total solid matter in the wastewater is mostly organic. But only large operations can justify the cost of TOC analysis.

Residue in wastewater. The amount of residue in the water affects BOD_5, COD, and TOC values. Operators may measure the amount of solids after the water is evaporated (total solids), the volatile solids (organic), or fixed solids (ash).

Settleable solids settle to the bottom of the wastewater sample in 1 hour. They are usually measured in a graduated Imhoff cone. Settleable solids show how much solid matter will settle in treatment systems in clarifiers and settling ponds. This test is easy to do, even at field sites.

To measure *total suspended solids,* or nonfilterable residue, a technician filters a measured volume of wastewater through a membrane filter (or glass fiber mat). The result is the dry weight of the total suspended solids (TSS) after 1 hour at 103 to 105°C (217–221°F).

After filtering, the technician evaporates the sample to find the weight of *total dissolved solids* (TDS), or filterable residue. Dissolved pollutants are difficult to remove from wastewater, so it is important to know they are there. The only way to remove TDS from wastewater is by using microorganisms, which absorb the dissolved solids so that they become part of the microbial cells, i.e., particulate matter.

Fats, oils, and grease (FOG) form an unattractive layer on the water. This layer does not allow oxygen from the air to dissolve in the water, so fish and other marine plants and animals cannot breathe. Heavy oil films clog the feathers of ducks and other water birds. Fats, oils, and grease also require a large amount of oxygen to break down.

Turbidity means the cloudiness of the water. Suspended organic matter, microorganisms, and other soil particles cause high turbidity. High turbidity levels do not necessarily mean that the water is polluted. Turbidity is measured using a candle turbidimeter. This measurement does not measure the amount of suspended matter, but it measures how the suspended matter affects light traveling through the water.

Waste material contains *nitrogen* in many forms, ranging from reduced ammonium to oxidized nitrate compounds. High concentrations of nitrogen can be toxic to animals and some plants. The most common forms of nitrogen found in wastewater are ammonia, proteins, nitrites, and nitrates. The total Kjeldahl nitrogen (TKN) method measures reduced forms of nitrogen, i.e., organic nitrogen and ammonia. Other tests measure the oxidized forms, i.e., nitrate and nitrite.

Wastewater contains *phosphorus* as orthophosphate and polyphosphate in organic or inorganic compounds. Water naturally contains trace amounts of dissolved phosphates, but too much can harm marine life. Trained technicians use chemical reagents and a colorimeter or spectrophotometer to measure phosphorus levels.

Wastewater may contain high levels of *sulfur* from sulfur dioxide or sodium bisulfide used to process fruits. Sulfides increase the BOD of the water and make drinking water smell and taste bad, so it is important to check how much sulfide is in wastewater. Trained technicians can easily measure sulfate and sulfide levels using basic equipment.

SOLID-WASTE DISPOSAL

Disposal of solid waste is a major challenge for the food industry. For example, by the year 2000, California will require every city to halve the amount of waste it sends to landfills. For food industries such as canneries, 65% of the raw materials delivered to the plant becomes solid waste. Food processors usually truck the wastes to a municipal garbage dump. If there is no dump nearby and the processor dumps wastes on the plant site, they soon smell bad and attract insects. Some processing firms compost solid wastes and use the compost as a fertilizer.

During composting, microbes break down the organic matter in the waste to make humus. Humus is very fertile dirt and is easy to till. Composting has four steps:

1. Solid-waste material is comminuted (ground) so that the microbes can reach the organic matter.
2. The comminuted waste is stacked in bins about 2 meters (2 yards) high and 3 meters (3 yards) wide.
3. The waste is aerated.
4. The compost is comminuted again.

Processors may add bacteria to speed up the composting process. Aerobic thermophilic microorganisms found naturally in the waste material can make compost in 10 to 20 days, depending on the temperature and type of waste.

Some food product wastes can be dried and ground for animal feed. Examples are the liquid waste from tomato processing, the residue from alcohol manufacture, citrus wastes, processed whey, and rendered animal by-products.

LIQUID-WASTE DISPOSAL

Wastewater can be recycled by removing the solids. The amount of wastewater that can be conserved and salvaged depends on the:

- Availability of treatment facilities
- Cost
- Market value of materials that are recovered from the wastewater
- Local regulations about wastewater quality
- Surcharges for plants discharging into public sewers
- Amount of wastewater

Economics often decide how much solid waste, waste concentrates, and blood (from meat processing) are sent to the sewer.

Whenever food processors handle, process, package, and store food, they generate wastewater. The amount of wastewater, the concentrations of pollutants, and the type of waste in the water have economic and environmental effects.

Cleaning compounds and sanitizers can increase the BOD and COD because they contain surfactants, chelators, and polymers, as well as organic acids and alkalis. Conveyor lubricants contain similar chemicals that increase the BOD and COD of the wastewater. However, these compounds make up less than 10% of the BOD and COD in wastewater from food-processing plants. Water used for sanitation from a food-processing plant could make up 30% of the total wastewater. Because the compounds do not increase the BOD and COD very much, the main concern is their effects on pH.

It is often cheaper for food processors to spend money on preventing and using waste products than on waste treatment facilities. But many food plants cause pollution because of their waste products. Municipal waste treatment plants may not have the capacity to treat this waste, so many food plants need special waste facilities. Wastewater treatment is a developing technology that needs cooperation between the Environmental Protection Agency (EPA), suppliers, and processors.

Pretreatment

Regulations usually require food processors to pretreat wastewater before discharging it into the municipal waste treatment system. The local sewer use ordinance limits the amounts of various wastes in wastewater. According to the EPA, many wastewaters from processing plants are acceptable and biodegradable.

Municipal sewage plants normally restrict wastewater discharge from food-processing plants. Food-processing waste does not usually contain toxic substances, but the municipal sewage plant cannot treat certain wastes, or the waste may cause an obstruction or damage to the facility. Difficult wastes include oils and fats, plant and animal tissues, and plant and animal waste (feces). Therefore, food processors need to isolate and pretreat the waste stream before sending it to the municipal waste treatment facility. The food processor needs to calculate whether it will be cheaper to pretreat the waste or pay a surcharge to the municipal service so that it can expand its system to cope with the extra waste in the water.

Surcharges are based on the flow rate, BOD, and amounts of suspended solids and grease. It is usually cheaper for smaller plants to pretreat wastewater just enough to make sure that it meets municipal regulations. Larger processors often find that it is better to pretreat wastewater more than the ordinance requires. Some plants pretreat the water enough that the municipal system reduces their surcharges. Many large-volume processors treat all of their wastewater to avoid high surcharges or because the municipal plant cannot handle the extra waste.

Pretreating wastewater beyond the level required by the local ordinance has several advantages:

- Grease and solid materials that the processor recovers from plant and animal waste products have a good market value. Soap plants, feed plants, and other industries will often buy these products at profitable prices.
- Additional pretreatment can reduce municipal charges and surcharges.
- Additional pretreatment reduces complaints from the municipal treatment facility.

The following disadvantages may discourage food processors from pretreating wastewater:

- Pretreatment facilities are expensive and make the processing operation more complex.
- Wastewater treatment can be expensive to maintain, monitor, and document.
- Pretreatment facilities are taxed along with other property unless state regulations allow tax-free waste treatment.

If the processor pretreats wastewater, she or he should base the process on findings from the waste-disposal survey. The processor needs to know about the waste the plant produces and any waste conservation and water reuse systems that are available to help design and estimate the cost of a pretreatment system.

Most pretreatment processes equalize waste flow and separate matter that floats and solids that settle. Processors may add lime and alum, ferric chloride ($FeCl_3$), or a selected polymer to help separation. The system may use paddles to help coagulate the suspended solids. Vibrating, rotary, or static screens concentrate the separated floating and settled solids before they are separated.

Flow equalization. Flow equalization and neutralization even out the load in the waste stream. The system needs a holding tank and pumping equipment so that the amount of wastewater that the plant releases does not fluctuate. This step reduces costs, whether the processing firm treats its own wastewater or discharges it into the municipal system after pretreatment. The equalizing tank can store wastewater before it is recycled, reused, or released in a steady stream to the treatment facility around the clock. The flow into the tank varies, but the flow from the tank is constant. Equalizing tanks can be lagoons, steel construction tanks, or concrete tanks, often without a cover.

Screening. Pretreatment often includes screening, using vibrating screens, static screens, or a rotary screen. Processors use vibrating and rotary screens more often because they allow pretreatment of more wastewater containing more organic matter. These screens allow water to keep flowing forwards, with solids constantly removed from the screen. Mesh sizes range from about 12.5 mm (0.5 in) in diameter for a static screen to about 0.15 mm (0.006 in) in diameter for high-speed circular vibrating screens. The system may use two screens together (e.g., static prescreen and circular vibrating screen) for efficient removal of solids.

Skimming. Processors use skimming to remove large, floating solids. The skimming equipment collects solids and transfers them to a disposal unit.

Primary Treatment

Primary treatment removes particles from the wastewater by sedimentation and flotation.

Sedimentation. Sewage contains a large amount of settleable solids. Screening and sedimentation can remove up to 40 to 60% of the solids, or about 25 to 35% of the BOD_5 load. Sedimentation removes some refractory (inert) solids that do not add to the BOD value.

Primary treatment equipment involves either a rectangular settling tank or a circular tank clarifier. Many settling tanks include slowly rotating collectors with attached paddles that scrape settled sludge from the bottom of the tank and skim floating scum from the surface.

The temperature of the wastewater affects sedimentation because heat causes convection currents and may stop some particles from settling. Eliminating the surface scum removes grease.

Flotation. Floatation removes oil, grease, and other suspended matter from wastewater. Flotation is useful in the food industry because it removes oil from wastewater.

Dissolved air flotation (DAF) removes suspended matter from wastewater using small air bubbles. Recycled, clarified wastewater is pressurized by air injection in a pressure tank. This pressurized water combines with raw wastewater and flows into the clarification vessel. The release of pressure causes tiny air bubbles to form. When particles attach to the air bubbles, the specific gravity of the particle becomes less than that of water. The bubbles rise up to the surface of the water, carrying the suspended particles with them.

Other ways to create air bubbles in wastewater include use of rotating impellers or air diffusers that form air bubbles at atmospheric pressure, saturation of the liquid with air and then placing it in a vacuum to create bubbles, or saturation of the liquid with air under high pressure.

Processors often use flocculating agents to pretreat wastewater before treatment in a DAF unit. Treatment by DAF is common because it is fast and removes solids with similar or lower densities than water. This treatment technique is expensive to install and run.

DAF systems keep bacteria alive inside the system to biodegrade pollutants in the wastewater. Food processors have also adapted floatation technology to handle sludge and for secondary and tertiary treatments. Flotation is popular with food processors who have large amounts of grease and oil in their wastewater.

Sludge collected from primary treatment contains about 2 to 6% solids and needs to be concentrated before final disposal. The cost of treating and disposing of sludge is the most expensive part of sewage treatment, unless sludge is used as a fertilizer or for some other practical purpose. Some treatment systems biodegrade most of the organic matter and leave very little sludge. These systems can lower the cost of treatment and disposal. If processors can sell the sludge or use it to make another product, disposal costs are lower. The salvaged material can make enough profit to cover other treatment costs. Processors can also use biological oxidations to treat recovered solids (sludges) before ultimate disposal.

A new method uses a series of coagulants made from corn starch to separate oil, grease, and suspended solids from wastewater. The processor can purify the grease and solids recovered from the DAF. The starch-based coagulants are added to an equalization tank before the DAF system. They reduce the surface charge on the solids and grease, so that they coalesce together and are more easily removed by DAF.

Secondary Treatment

Secondary treatment usually involves biological (or bacterial) oxidation of dissolved organic matter. Secondary treatment may use lagoons or sophisticated activated sludge processes. It may also include chemical treatment to remove phosphorus and nitrogen or to help solids clump together.

Although primary treatment removes screenable and readily settleable solids, the wastewater still contains dissolved solids. The main purpose of secondary treatment is to continue removing organic matter and produce water with a low BOD and with very little suspended solids. Microbial flora can convert some of the dissolved solids into final oxidation products (such as carbon dioxide and water) or into removable material. These reactions need oxygen. After treatment, gravity sedimentation separates the microbial suspended solids from the water.

Both primary and secondary treatments produce waste sludges. These sludges usually need more treatment before final disposal. Anaerobic lagoons and aerobic lagoons, or stabilization ponds, treat wastewater and stabilize sludge. This treatment technique is popular because it is cheap and easy to build and operate. Anaerobic and aerobic lagoons do not work well for extremely large waste loads and are not a good choice if land costs are very high.

Anaerobic lagoons. The depth of anaerobic lagoons varies from 2.5 to 3.0 meters (2.5–3.0 yards). The ratio of surface area to volume should be as low as possible (i.e., the amount of surface should be as small as possible for the volume of the lagoon). Heavy organic loads create anaerobic conditions (i.e., no air) throughout the entire lagoon. Under anaerobic conditions, anaerobic microbes digest the organic matter. The temperature must be 22°C (72°F) or higher for 4 to 20 days. The lagoon reduces the BOD of the waste by 60 to 80%, depending on the BOD of the original waste and the amount of time it spends in the lagoon. Anaerobic lagoons work for primary treatment or secondary treatment of wastewater with high organic loads or for treatment of sludge. The products of anaerobic lagoons usually go to aerobic lagoons or trickling filters because they are still high in organic matter.

Some treatment processes combine anaerobic and aerobic treatment. An anaerobic tank reactor breaks down complex organic compounds into carbon dioxide, methane, and simple organic compounds. The anaerobic tank reduces BOD_5 by 85 to 95%. The gases separate from the water and contain about 65 to 70% methane. The wastewater flows on to an aerobic reactor for further treatment.

A combination of anaerobic and aerobic treatment can handle wide variations in the amount of wastewater. Anaerobic treatment responds slowly to changes in flow because anaerobic microorganisms grow slowly, but the faster-growing aerobic microorganisms can soon handle higher loads in the water after the anaerobic treatment.

Aerobic lagoons. Aerobic lagoons use mechanical aerators to add atmospheric oxygen to the wastewater. The two types of aerated lagoons are (1) aerated facultative lagoons, which mix the waste enough to dissolve oxygen but not enough to keep all the solids suspended, or (2) completely mixed aerated lagoons, which are mixed enough to keep all solids suspended. About 20% of the BOD sent to an aerobic lagoon is made into sludge solids, and the lagoon reduces the BOD of the water by 70 to 90%.

Trickling Filters

In trickling filters, a thin layer of wastewater trickles over stationary rocks arranged above a drain. Exposing large surface areas of wastewater to the atmosphere aerates the wastewater. Bacterial action and biological oxidation reduce BOD and suspended solids (SS). Bacteria grow on and attach to the rock surfaces. The trickling filter is often called a *fixed-bed system.*

Activated sludge. Processors often use the activated sludge process to treat wastewater. This process requires an aeration tank or basin, a clarifier, and a pump to return some of the settled sludge to the reactor and send the rest into the waste stream. A portion of the settled sludge is returned to mix with wastewater entering the reactor. The returned sludge is called *activated sludge* because it contains microorganisms that actively break down the waste. The activated sludge process is often called a *fluid-bed biological oxidation system.*

A conventional activated sludge system is designed for continuous secondary treatment of domestic sewage. It cannot treat inorganic dissolved solids, but it removes all organic matter in the wastewater. This process uses either surface aerators or air diffusers to mix the wastewater and sludge. Aeration equipment is fairly expensive, and electrical costs are high. This treatment process can be very efficient (95–98%) and can be modified to remove nitrogen and phosphorus without using chemicals.

Oxidation ditch. This treatment technique is an efficient, easy-to-operate, and economical process for treating wastewater. The process keeps waste materials in contact with activated sludge for 20 to 30 hours with constant mixing and aeration. The temperature influences how well the oxidation ditch removes waste from water. During cold weather, biological particles may clump together so that the system is less efficient.

The typical oxidation ditch is either a single or multiple closed-loop channel(s). Oxidation ditches need very little operator attention once they are working properly. Several food-processing firms use oxidation ditches for wastewater treatment.

Processors are looking at the new total-barrier oxidation ditch (TBOD) as a way to treat municipal food-processing and industrial wastewater. The TBOD biologically purifies water as it mixes oxygen with waste particles and allows the bacteria to feed on the pollutants. A constant, powerful flow of wastewater prevents settling at the bottom of the ditch reactor. The aeration and pumping system consists of underwater turbine draft tube aerators that transfer oxygen into the liquid.

Land application. The two types of land application techniques that work best are *infiltration* and *overland flow.* If land application techniques are not run properly, pollutants can harm plants, soil, and surface and ground waters. However, both of these techniques can remove organic carbon from wastewater. The infiltration flow system removes about 98%, and the overland flow system removes about 84% of organic matter. The overland flow system usually causes less pollution of drinkable groundwater supplies.

Land application now has limited use for food-processing wastes. Minerals and other materials can build up in the soil, and the residues can have long-term effects on plants.

Rotating biological contactor. The rotating biological contactor (RBC) system is similar to trickling filters. The initial cost of the equipment is high, but the operating costs are moderate. RBC units have a number of large-diameter (about 3 meters or 3 yards), lightweight discs mounted 2 to 3 cm (about 1 in) apart on a horizontal shaft. The discs are partly (30–40%) immersed and rotate slowly (0.5–10 rpm) as wastewater passes through a horizontal open tank, which usually has a semicircular base to fit the shape of the discs.

The RBC unit has microorganisms attached to the surface of the discs that grow on nutrients from the wastewater. Rotating the disc surface so that it is sometimes above the water aerates the microorganisms and the thin film of water on the disc's surface. Faster rotation speeds increase the amount of dissolved oxygen in the tank.

Tertiary Treatment

Tertiary treatment processes for wastewater are known as *advanced wastewater treatment.* They improve the quality of treated wastewater to meet regulatory guidelines. Tertiary waste treatment removes pollutants such as colors, smells, brines, and flavors from food-processing wastewater.

Physical separation. Sand filters and microstrainers purify treated wastewater. Both of these methods remove suspended solids as small as micrometers in diameter.

The *microstrainer* is a rotating cylinder covered with a screen (fine mesh nylon or metal fabric) that lays in an open tank. Wastewater enters the cylinder and filters out through the screen. The cylinder rotates slowly, and the exposed section above the wastewater surface is backwashed to clean the screen and collect the solids into a separate channel. The size of particles removed by microstraining depends on the size of the screen pores. This tertiary treatment is relatively cheap, because the screens clean themselves and running and maintenance costs are low. Over time, the screen becomes partly clogged, making the system less efficient. Eventually the system needs a new screen. Also, microorganisms can grow inside the cylinder and create slime on the screen. Ultraviolet light or chlorine can reduce slime formation.

Tertiary lagoons. These maturation lagoons or *polishing ponds* treat secondary wastewater from activated sludge or trickling filter systems. The lagoons are usually 0.3 to 1.5 meters (0.3–1.5 yd) deep. Natural aeration, mechanical aeration, or photosynthesis supply oxygen to the lagoon. This system can reduce BOD and SS by up to 80 to 90%. Temperature affects how efficiently this system removes waste. This simple treatment method needs almost no attention to equipment or power during day-to-day operation. However, the lagoons need more land than any other treatment process.

Chemical oxidations. Various chemicals can oxidize wastewater components during tertiary treatment. One such chemical is *ozone*. Equipment to generate ozone at a reasonable cost is now available. Ozone breaks down in water to form oxygen and nascent oxygen, which rapidly reacts with organic matter. This process also disinfects the water, removes its taste and odor, and bleaches it. Other chemicals that oxidize wastewater are chlorine, chlorine dioxide, oxygen, and permanganate.

Disinfection

To protect public health, processors should disinfect treated wastewater before finally releasing it. The number of microbes in the wastewater decreases during primary and secondary treatment. Pathogenic microorganisms die after long periods of time in natural environments. Disinfectants can react with organic matter, so it is better to disinfect at the end of wastewater treatment. Table 8.2 shows the usual numbers of microbes in domestic wastewater.

TABLE 8.2. Microbial Characteristics of Domestic Wastewater

Microorganism	Quantity per 100 ml Wastewater
Total bacteria	$10^9–10^{10}$
Coliforms	$10^6–10^9$
Fecal *Streptococci*	$10^5–10^6$
Salmonella typhosa	$10^1–10^4$
Viruses (plaque-forming units)	$10^2–10^4$

Source: Arceivala, S. J. 1981. *Wastewater Treatment and Disposal.* Marcel Dekker, New York.

Chemical oxidants, irradiation (ultraviolet, gamma, and microwave), and physical methods (e.g., ultrasonic disruption and heat) can all disinfect water.

Chlorination is now less popular because organohalides in chlorinated water may cause cancer. Too much chlorine in wastewater can also be toxic to fish. Chlorination and other chemical treatments do not kill all microorganisms. Some algae, spore-forming bacteria, and viruses (including pathogenic viruses) survive chlorination.

Ultraviolet irradiation equipment can disinfect moderate amounts of water and has no effects on plants, animals, or other organisms in the water stream. Heat treatment works well but is not practical for large amounts of water.

SUMMARY

- A survey of the type and amount of waste and the amount of water used in a food-processing plant can help decide on the best waste treatment system.
- BOD, COD, DO, TOC, SS, total suspended solids (TSS), TDS, and FOG are all tests that measure the amount of waste pollution.
- Food processors can recycle wastewater. They can recover and reuse solids in wastewater.
- The basic phases of wastewater treatment are pretreatment (flow equalization, screening, and skimming), primary treatment (sedimentation and flotation), secondary treatment (anaerobic lagoons, aerobic lagoons, trickling filters, activated sludge, oxidation ditch processes, land application, and rotating biological contactors), and tertiary treatment (physical separation, tertiary lagoons, and chemical oxidation).
- Processors should disinfect treated wastewater after the other treatment phases so that organic matter does not react with the disinfectant.

BIBLIOGRAPHY

AOAC. 1990. *Official Methods of Analysis,* 15th ed. Association of Official Analytical Chemists, Washington, D.C.

Burleson, G. R., Murray, T. M., and Pollard, M. 1975. Inactivation of viruses and bacteria by ozone, and without sonication. *Appl. Microbiol.* 29:340.

Carawan, R. E., Chambers, J. V., Zall, R. R., and Wilkowske, R. H. 1979. *Meat Processing Water and Wastewater Management.* Extension Special Report AM-18C. North Carolina State University, Raleigh.

Culp, R. L., Wesner, G. M., and Culp, G. L. 1978. *Handbook of Advanced Wastewater Treatment.* Van Nostrand Reinhold, New York.

Green, H. J., and Kramer, A. 1979. *Food Processing Waste Management.* AVI Publishing Co., Westport, Conn.

Marriott, N. G. 1994. *Principles of Food Sanitation,* 3d ed. Chapman & Hall, New York.

National Canners Association. 1971. *Liquid Wastes from Canning and Freezing Fruits and Vegetables.* Off. Res. Monitoring, U.S. Environmental Protection Agency, Washington, D.C.

Sofranec, D. 1991. Wastewater woes. *Meat Proc.* Nov. 1991, p. 46.

STUDY QUESTIONS

1. Why must food processors deal with waste immediately?

2. What happens when untreated organic waste pollutes rivers or lakes?

3. What is BOD?

4. What are the three steps in a wastewater survey?

5. Name five tests for the amount of pollution in wastewater.

6. Give three ways food processors dispose of solid waste.

7. What are the advantages and disadvantages for food processors who pretreat their own waste?

8. What are the purposes of pretreatment, primary treatment, secondary treatment, tertiary treatment, and disinfection of wastewater?

TO FIND OUT MORE ABOUT WASTE DISPOSAL

1. Contact your local municipal water treatment facility. What types of system do they use to treat and disinfect the water? Does the area have any particular water treatment problems? What local industries add to the burden on the water treatment system? What kinds of waste do these industries produce?

2. Contact a large food-processing facility, in your area if possible. Ask them whether they process their own waste completely or partially. If they process their own waste, what systems do they use? Why did they decide to use these systems?

3. Contact the American Water Works Association, (303) 795-2449, and request a copy of their *Buyers' Guide and Publications Catalog.*

CHAPTER 9

Pest Control

ABOUT THIS CHAPTER

In this chapter you will learn:

1. Which pests (insects, rodents, and birds) cause the most problems in food establishments
2. How to prevent pest infestations through inspections and good housekeeping
3. How to detect pest infestations
4. Why pest control is so difficult but essential
5. How to control pests, using chemicals, physical and mechanical methods, biological methods, and integrated pest management
6. How to use pesticides safely

INTRODUCTION

This chapter will not make you an expert on pest control, but it will give you a basic understanding of how insects, rodents, and birds contaminate food supplies. Foodservice staff need to know about pests that can contaminate the food supply and how to control them. The number of species of insects, rodents, and birds that cause problems in the food supply is small, but they cost the food industry billions of dollars every year.

Food establishments may employ an outside exterminator. Also, one or more employees (depending on the size of the organization) may be trained to take care of pest control.

Tidiness and cleanliness help keep pests away from the facility. They make it easier to exterminate pests, more difficult for pests to get into the building, and harder for pests to find a place to live and reproduce. Insects and rodents will not be able to find shelter or food if rubbish, food waste, and equipment that is no longer used are disposed of quickly. Enclosed areas under shelves, platforms, chutes, and ducts, and holes in walls and insulation, make good homes for pests, especially if debris collects in these areas.

INSECTS

Cockroaches

Cockroaches are the most common pests in food-processing plants and foodservice facilities throughout the world. These pests must be controlled because they carry and spread various diseases. Many carry the microorganisms that cause poliomyelitis and cholera.

Cockroaches spread microorganisms when they crawl on, bite, and chew food. They prefer high-carbohydrate (starchy or sweet) foods, but will eat any food, human waste, decaying food scraps, dead insects (including other cockroaches), shoe linings, paper, or wood. Cockroaches are most active in dark places and at night where humans are less likely to disturb them.

Each month, these pests lay small egg cases containing 15 to 40 eggs. Some cockroaches deposit the egg case in a hiding place to protect it. Shortly after they hatch, young cockroaches begin eating the same things the adults eat. Young cockroaches look like adults, except that they are smaller and do not have wings. They develop wings after they grow and shed their skin several times. Cockroaches may live over a year and mate several times.

Food sanitarians need to know the type of cockroach that has infested an establishment, so that they can use the right method to remove it. The three most common cockroach species in commercial establishments in the United States are the German cockroach, the American cockroach, and the Oriental cockroach. However, other cockroaches, such as the field cockroach (found on crops), are starting to cause problems in homes and food establishments in parts of the southern United States.

Species

German cockroach. The German cockroach is pale brown, with two dark-brown stripes behind the head. All adults have well-developed wings. The female carries the egg case behind the tip of her abdomen until the eggs hatch.

In food establishments, German cockroaches can infest the main processing or preparation rooms, storage areas, offices, and washroom facilities. They prefer to live in warm crevices near warm pipes, stoves, boilers, or other heating equipment, and do not usually live in basements. German cockroaches are common in restaurants and may be found anywhere in a room, from the floor to the ceiling.

American cockroach. This is the largest cockroach in the United States. Adults are reddish-brown, and the young are pale brown. The female hides the egg cases as soon as she produces them.

American cockroaches live in open, wet areas such as basements, sewers, drainage areas, and garbage areas, as well as storage rooms and loading docks. They prefer cool places that have larger crevices than the German cockroach prefers.

Oriental cockroach. The Oriental cockroach is shiny dark brown or black. The male has very short wings, and the female has no wings. Its young are pale brown. The female hides the egg cases soon after she produces them.

This species prefers similar areas to those preferred by the American cockroach. In food plants, they normally live in basements or damp storage areas.

Detection

Cockroaches live anywhere where food is processed, stored, prepared, or served. The insects hide and lay eggs in dark, warm, hard-to-clean areas. They like small spaces in and between equipment and under shelf liners. When cockroaches need food that is not in these areas or they are forced out by other cockroaches, they will come out into the light.

One of the easiest methods of checking for cockroaches is to walk into a dark production or storage area and then turn on the lights. Cockroaches produce a strong, oily odor from their glands. They also leave feces (small, black or brown, and spherical) everywhere they go.

Control

Cockroaches infest food establishments all year round, so all workers need to continuously use good sanitation and chemicals to keep them under control. Good sanitation is the most important form of control. Because these insects will eat almost anything, workers should keep all areas of the operation, including washroom facilities, tidy and clean.

Filling cracks in floors and walls with caulking or other sealants reduces the number of cockroaches. Spaces between large pieces of equipment and their bases or the floor should be sealed. These spaces provide an ideal home for pests. Cockroaches get into food establishments as cockroaches or as eggs in boxes, bags, raw foods, or other supplies. All deliveries should be examined with the removal of any insects or eggs. Cartons and boxes should be taken to the trash area as soon as the supplies are unpacked.

An outside pest controller may handle chemical control of pests, but a combined chemical and sanitation program within the organization can be more effective and cheaper. Cockroach control uses baits and bait stations, fungi, and sometimes nematodes (parasitic worms). Cockroaches become inactive at about 5°C (41°F), so refrigeration and freezing reduces their numbers.

The chemical amidinohydrozone can kill cockroaches that resist other poisons. An insecticide such as diazinon sprayed into hiding places works well if the pests have not developed a resistance to this compound. Pyrethrin-based insecticide is sometimes added to diazinon to force the insects out of hidden areas into the sprayed area. Some form of diazinon can be used in spots, cracks, or crevices, but not in food-handling areas. Workers must follow the directions on the label when they use any insecticide.

Other Insects

The most common seasonal insects in foodservice and food-processing plants are houseflies and fruit flies.

Housefly. It lives all over the world and is an even greater health problem than the cockroach. It carries several types of pathogens to humans and their food, causing typhoid, dysentery, infantile diarrhea, streptococcal and staphylococcal infections, and other diseases.

Flies eat animal and human wastes and collect pathogens from the waste on their feet and wings and in their mouths and guts. When flies walk on food or eat food, they leave pathogens on it. Flies take in food as liquid, so they secrete saliva onto solid food to liquefy it before they suck it up. Fly saliva is loaded with bacteria that contaminate food and equipment.

Control of flies is challenging, because these pests can get into buildings through openings only slightly larger than the head of a pin. Flies normally stay close to the area where they hatch, but they are attracted to smells and decaying materials. Air currents often carry flies further than they would fly. Flies prefer to live in warm areas out of the wind, such as garbage can rims. Warm, damp, shaded decaying material provides an ideal place for housefly eggs to hatch and grow as fly larvae or maggots.

There are more houseflies during late summer and fall because the population builds rapidly during warm weather. Once adult flies enter buildings for food and shelter, they usually stay there. Flies are most active at temperatures between 12°C and 35°C (54–95°F). Below 6°C (43°F) they are very inactive, and below 5°C (41°F) they die within a few hours. Heat paralyzes flies at about 40°C (104°F) and kills them at 49°C (120°F).

The size of a housefly population is difficult to control, because they often breed away from food establishments in decaying material. The best ways to control the fly population are to prevent flies from entering processing, storage, preparation, and serving areas and to reduce their numbers within these areas.

Sanitarians can prevent flies from getting into food establishments by removing waste materials from food areas promptly and thoroughly. Air screens, mesh screens (at least 16 mesh is recommended by the U.S. Public Health Service), double doors, and self-closing doors help keep flies out. Staff should keep doors closed whenever possible and open them for the shortest possible amount of time for receiving or shipping.

To keep flies away from a food establishment, staff should store the outdoor garbage as far away from doors as possible. If the facility stores garbage inside, a wall should separate this area from other areas, and refrigeration should be used to reduce decay and fly activity. Garbage containers should have lids.

If flies do get into the facility, an electric fly trap attracts the adults to a blue light and kills them using an electric grid. Electric fly traps should run all day, and the catch basin should be cleaned daily. Aerosols, sprays, or fogs of chemicals such as pyrethrins can help control flies. But the results are temporary, and chemicals are restricted in food facilities. Therefore, keeping flies out and using fly traps are preferred methods of controlling flies.

Fruit fly. The fruit fly is smaller than the housefly and is also most abundant in late summer and fall. Adult fruit flies have red eyes and light-brown bodies. They are attracted to fruit, especially rotting fruit. These pests are not attracted to sewage or animal waste, so they carry fewer harmful bacteria.

The life cycle and feeding habits of fruit flies are similar to those of houseflies, except that these insects are attracted specifically to fruits. Fruit flies grow fastest in late summer and early fall when there are plenty of rotting plants and fruits. Fruit flies live about 1 month.

It is hard to get rid of fruit flies completely. Mesh screens and air screens keep them out of food establishments. When they do get in, electric traps help keep their numbers down. The best way to control these pests is to get rid of rotting fruits and fermenting foods quickly.

Other insect pests. In food-processing and foodservice operations, they include ants, beetles, and moths. These pests leave webs or holes in food and packaging materials, and they can be controlled by keeping the facility tidy and well ventilated, keeping storage areas cool and dry, and rotating stock.

Ants often build nests in walls, especially around heat sources, such as hot water pipes. Sponges saturated with syrup can act as bait to find out where ants are living and where to place insecticide. Because ants, beetles, and moths need only very small amounts of food, good sanitation and proper storage of food and supplies are essential to keep these pests away.

Silverfish and firebrats can live in cracks, baseboards, window and door frames, and between layers of pipe insulation. Because these pests like undisturbed areas, thorough and frequent cleaning usually keeps them away. Silverfish prefer a moist environment, e.g., basements and drains. The firebrat likes warmer places, such as around steam pipes and furnaces.

INSECT DESTRUCTION

Pesticides

Pests should be destroyed without using chemicals whenever possible, because pesticides can be dangerous. But sometimes pesticides are necessary if other methods do not work. Food producers should consult a professional pest control firm if they do not know how to apply pesticides properly. Only commercial pest controllers are allowed to apply restricted pesticides. Even when an exterminating firm is used, supervisors at the food establishment need a basic understanding of pests, insecticides, and regulations concerning use of chemicals.

Residual insecticides. These insecticides are effective for an extended period of time. Some residual insecticides are illegal in food areas. It is important not to contaminate food, equipment, utensils, and supplies that people will eat and workers will touch. Staff who use these chemicals should understand the terms on the product labels, how the product is allowed to be used, and effects the product may have.

Operators apply residual insecticides three ways:

- *Mass application* to floors, walls, and ceilings.
- *Spot treatment* in cracks and crevices where insects are likely to live. Spot areas may be on floors and walls or around equipment and supplies.
- *Crack and crevice treatment* for areas where insects hide or enter buildings, for example, expansion joints in building construction and between equipment and floors. It is important to treat these places, because the openings often lead to hollow walls or equipment legs and bases where pests like to live. Staff should also treat conduits, junction or switch boxes, and motor housings.

Nonresidual insecticides. These insecticides control insects only at the time they are applied. Operators apply nonresidual insecticides in two ways:

- *Contact treatment* means applying a liquid spray for an immediate insecticidal effect. Operators should use only this method when the spray is likely to touch the pests.
- *Space treatment* uses foggers, vapor dispensers, or aerosol devices to disperse insecticides in the air. This technique can control flying insects and crawling insects in the exposed area. Space spraying helps control the number of insects.

Operators may use nonresidual insecticides in food production areas only when food is not exposed. Examples of nonresidual insecticides are pyrethrins and pyrethroids. Aerosols are good for killing flying and exposed insects and are often released using a timer at a time when food production and contact does not occur.

The food industry uses fumigants to control insects that attack stored products. Fumigants are good at reaching hidden pests and are often used for space treatment on weekends when the plant is not operating, so that food and workers are not exposed. Ventilation machinery or fans help disperse fumigants properly.

The following chemicals are common fumigants for insects:

Phosphine: This insecticide comes in a permeable package or in pellets. The gas is very flammable, so it is important to follow the manufacturer's instructions for use and storage.

Methyl bromide: This nonflammable fumigant is widely used.

Ethylene oxide: Operators often mix this fumigant with carbon dioxide in a ratio of 1:9 (by weight) to reduce flammability and explosiveness. A professional pest controller has to apply this compound.

Other Chemicals for Controlling Insects

Baits use sugar or other foods to attract insects and an insecticide to kill them. Baits may be less convenient than other methods, but they can work well against ants and cockroaches in inaccessible areas and against outside fly populations. Baits are a poisonous food, so it is important to use and store them carefully. Operators may scatter commercial dry granular baits over feeding surfaces every day, or as needed, to control insect populations. Granular fly baits work well outdoors. Liquid baits contain an insecticide in water with sugar, corn syrup, or molasses. Operators spray or sprinkle liquid bait on walls, ceilings, or floors where flies are found. They should use fly bait regularly during the summer months to control growth of insect populations.

Mechanical Methods

Mechanical devices to control insects do not work very well. Fly swatters spread insect carcasses and bacteria, so they should not be used in food-processing, storage, preparation, or sales areas. The best mechanical device for controlling insects is an air curtain. This prevents insects and dust from entering food establishments and also reduces cold-air loss from refrigerated facilities. Air curtains can be used even at large entrances used by loading trucks or large equipment. An air curtain is created by a fan that sweeps air down over door openings and is usually mounted outside and above door openings.

Insect light traps. One of the safest and best ways to control flies is to use insect light traps. The advantage of this technique is that it is not toxic.

Insect light traps conduct a high-voltage, low-amperage current across a grid (see Fig. 9.1). This creates a light source that attracts flies, but when they fly toward it, they are electrocuted. Some light traps have a "black light," which works at night, and a "blue light," which works in the daytime.

Managers should install insect light traps in food-processing plants and warehouses in the following stages:

Stage 1, interior perimeter: Place these units near shipping and receiving doors, employee entrances, and staff doors that lead to the outside or anywhere else where

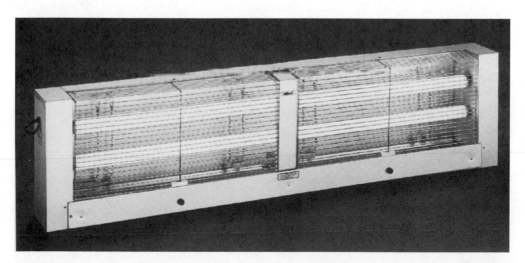

FIGURE 9.1. An insect light trap that attracts flies to the light source, subsequently electrocuting them. (*Courtesy of Dan Gilbert Industries, Inc.*)

flying insects may enter. Install units 3 to 8 meters inside doors, away from strong air currents and out of traffic areas where forklifts or other equipment may damage them.

Stage 2, interior: Place these units along the path that insects may follow to food-processing areas. Use units with wings within processing areas to prevent dead insects from falling on the floor or on processing equipment.

Stage 3, exterior perimeter: Protect covered docks, especially if they are used to store refuse. Install units between the insects and the entrances, but not directly at the entrances.

It is important to replace the light source every year for it to work properly. Operators need to place traps carefully so that they attract as many insects as possible, but not insects from outside. Staff should empty the pan that collects the electrocuted insects regularly so as not to attract dermestid beetles and pests that feed on dead insects.

Sticky traps. Sticky traps consist of sticky flypaper or a flat piece of plastic covered with a slow-drying adhesive. Yellow plastic strips with a sticky covering catch a wide variety of flying insects.

Biological Control

Biological control is a common part of integrated pest management (IPM) programs (discussed near the end of this chapter). Growth regulators may interrupt the life cycle of insects and prevent reproduction of mosquitoes, fleas, and other insects. Milled diatomaceous earth contains sharp microscopic particles that pierce the insects' wax coating so that they lose moisture and die. If particles of the shell get into the insect's body cavity, they interfere with digestion, reproduction, and respiration.

Hydroprene is an insect growth regulator (IGR) used to control cockroaches. The Environmental Protection Agency has approved its use in areas where food is present because it is relatively nontoxic and is not a pesticide. An IGR can disrupt the normal growth and development of immature cockroaches by deforming their wings and making them sterile.

Pheromone traps. These traps attract insects using insect sex hormones (pheromones) and have a trapping chamber to catch the insects. Some have a plastic funnel leading into a chamber that contains an insecticide strip. New products contain microencapsulated pheromones that release chemicals slowly over a long period of time. Pheromones are now available for various species, including fruit flies. Some food aromas may attract insects even better than pheromones.

RODENTS

Rodents such as rats and mice are difficult to control because they have highly developed senses of hearing, touch, and smell. These pests can identify new or unfamiliar objects in their environment and then protect themselves against these changes. Figure 9.2 shows some of the incredible things mice can do.

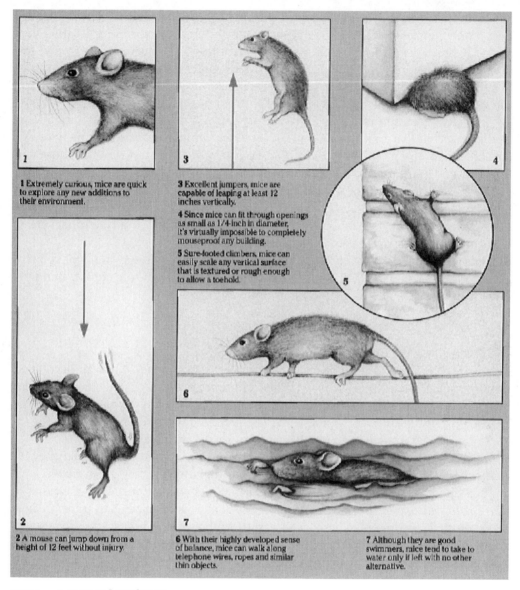

1 Extremely curious, mice are quick to explore any new additions to their environment.

2 A mouse can jump down from a height of 12 feet without injury.

3 Excellent jumpers, mice are capable of leaping at least 12 inches vertically.

4 Since mice can fit through openings as small as 1/4-inch in diameter, it's virtually impossible to completely mouseproof any building.

5 Sure-footed climbers, mice can easily scale any vertical surface that is textured or rough enough to allow a toehold.

6 With their highly developed sense of balance, mice can walk along telephone wires, ropes and similar thin objects.

7 Although they are good swimmers, mice tend to take to water only if left with no other alternative.

FIGURE 9.2. Mouse acrobatics.

Rats

Rats can enter buildings through openings as small as a quarter, climb vertical brick walls, and jump up to a meter (39 in) vertically and 1.2 meters (48 in) horizontally. These rodents are strong swimmers and can swim up through toilet bowl traps and floor drains. Rats have excellent balance and can even walk on suspended wires.

Rats are dangerous and destructive. The National Restaurant Association estimates that rodent damage costs $10 billion per year. This includes the costs of food they eat, food they contaminate, and structural damage to property, including damage from fires started when rats gnaw on electrical wires. The serious health hazard from contaminated food, equipment, and utensils is even more serious than economic losses. Rats transmit such diseases as leptospirosis, murine typhus, and salmonellosis. One rat dropping contains several million harmful microorganisms. When droppings dry out and fall apart or are crushed, air movement within a room can carry particles to food.

The most common rat in the United States is the Norway rat, sometimes known as the sewer rat, barn rat, brown rat, or wharf rat. Norway rats are normally brown and are 18 to 25 cm (7.5–10.5 in) long, excluding the tail, have a blunt nose and thick-set body, and tend to live in burrows.

Roof rats live in the South and along the Pacific Coast and Hawaii. They prefer high places and have better coordination than Norway rats. Roof rats are black or slate-gray and are 16.5 to 20 cm (7–8.5 in) long, excluding the tail. Roof rats create nests in trees, vines, and other places above the ground.

Rats and mice instinctively avoid open spaces, especially if they are light colored. Therefore, a 1.5-m-wide (5-ft-wide) band of white gravel or granite chips around the outside of a building helps keep rodents out.

Mice

Mice can enter a building through a hole as small as a nickel. Mice are also good swimmers, can swim through floor drains and toilet bowl traps, and have excellent balance. Like rats, mice are filthy rodents and spread many diseases. The house mouse is found everywhere in the United States and has a body length of 6 to 9 cm (2.5–4 in). It has a small head and feet and large prominent ears.

Mice do not need to drink water, because they can survive on the water they release when they metabolize food. But they do drink liquids if they are available.

It is easy to carry mice into food premises in crates and cartons. They are easier to trap than rats because they are less careful. Metal and wood-base snap traps are normally effective. Operators may space several traps about 1 m (3 ft) apart. Mice usually accept new objects, such as traps, often after about 10 minutes. Sodium fluorosilicate and chlorophacinone are poisonous tracking powders that work well to control mice.

Finding Rodent Pests

Rats and mice tend to be inactive during daylight hours, so they are not always obvious. Droppings are an obvious sign of rodent infestation. Fresh droppings are black, shiny, and pasty. Older droppings are brown and fall apart when touched. Figure 9.3 shows the features and droppings of Norway rats, roof rats, and house mice. Table 9.1 describes these three rodents.

Rats and mice usually follow the same path or runway between their nests and food sources. After awhile, grease and dirt from their bodies form visible streaks on floors and other surfaces. Rodents tend to stay close to vertical surfaces when they travel, so their tracks may go along walls, rafters, steps, or the sides of pipes. Rat and

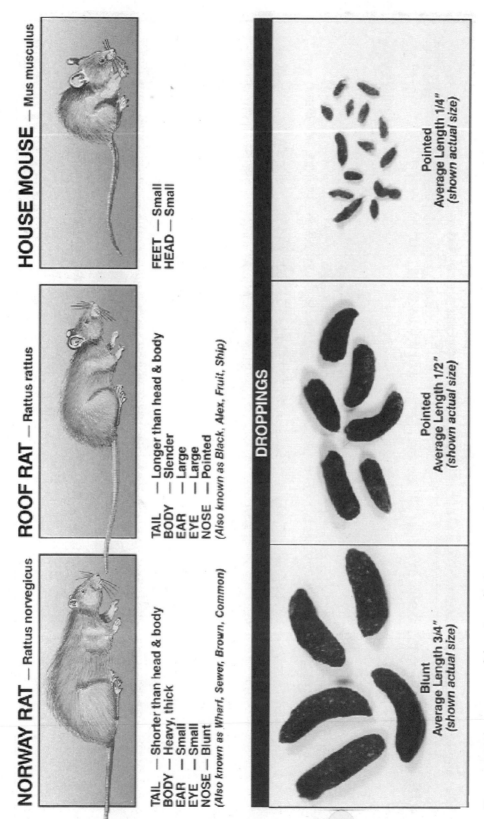

NORWAY RAT — Rattus norvegicus

TAIL — Shorter than head & body
BODY — Heavy, thick
EAR — Small
EYE — Small
NOSE — Blunt

(Also known as Wharf, Sewer, Brown, Common)

ROOF RAT — Rattus rattus

TAIL — Longer than head & body
BODY — Slender
EAR — Large
EYE — Large
NOSE — Pointed

(Also known as Black, Alex, Fruit, Ship)

HOUSE MOUSE — Mus musculus

FEET — Small
HEAD — Small

DROPPINGS

Blunt
Average Length 3/4"
(shown actual size)

Pointed
Average Length 1/2"
(shown actual size)

Pointed
Average Length 1/4"
(shown actual size)

FIGURE 9.3. Appearance and droppings of common rats and mice.

TABLE 9.1. Facts About Common Rats and Mice

Name	Average Weight, oz	Body	Tail	Ears	Color	Food Preferences	Food Consumed Per Day	Droppings Per Day	Urine Per Day, cc	Habits
Norway rat _Rattus Norvegicus_	10–17	Heavy, broad, 7–10 in long, blunt head	6–8.5 in long, lighter color on underside	Average, close to body	Brown to black on back and sides, gray to yellow-white on belly	Meats, fish, flour, fruits, vegetables;. eats most human foods	3/4–1 oz food 1/2–1 oz water No. young in litter, 6–18	30–180 # litters per year 3–7	15.7 Length adult life, 18 mo	Usually nests in basements & lower portions of buildings. Burrows in soil extensively. Active primarily at night. Fair climber, good swimmer.
Roof rat _Rattus rattus Alexandrinus_	6–12	Slender body 6.5–8 in long, pointed muzzle	7.5–10 in long. Uniform color, top and bottom	Large, prominent	Tawny back, grayish-white underparts	Seeds, fruits, vegetables, grain, eggs, etc.	1/2–1 oz food, up to 1 oz water No. young in litter, 6–14	30–180 # litters per year 3–7	15.7 Length adult life, 18 mo	Usually enters and nests in upper portions of buildings. May nest outside in trees (esp. palms), ivy, etc. Burrows very little. Excellent climber. Active at night.
House mouse _Mus musculus_	1/2–3/4	Average body 2.5–3.5 in long	3–4 in	Large, prominent	Dusky gray on back, lighter gray on belly	Meats, grains, cereals, seeds, fruits, vegetables; eats most types of human food	1/10 oz food, 1/20 oz water No. young in litter, 6	50 # litters per year 8	1.75 Length adult life, 15–18 mo	May nest in any portion of a building near food. Has an average range of 50 ft from nest. Active mostly at night. Nibbles small amounts of food frequently, rather than having large meals.

Used with permission, LiphaTech, Inc., Milwaukee, Wisconsin.

mouse tracks are easy to see on dusty surfaces. Operators can spread talc in areas where they think rodents are active to make tracks more visible. Long-wavelength UV light can show up urine stains as a yellow fluorescence on burlap bags, and a pale, blue-white fluorescence on kraft paper.

Gnaw marks also show when rats are present. The incisor teeth of rats are strong enough to gnaw through metal pipes, unhardened concrete, sacks, wood, and corrugated materials to reach food. Teeth marks are visible if gnawings are recent. Bumping noise, shrill squeaks, fighting noises, or gnawing sounds at night are clues that rodents are living in the building.

Control

Control of rodents, especially rats, is difficult because they adapt to the environment so well. The best way to control rodents is by proper sanitation. If entrances are sealed off, these pests cannot get in, and if there is no debris to eat, they will not survive. If sanitation is not good, poisons and traps will only reduce the rodent population temporarily.

Prevention of entry. To ratproof an establishment, staff need to block all possible entrances. Poorly fitting doors and poor masonry around external pipes need flashing or metal or concrete covers or fillings. Vents, drains, and windows need screens. Decay in building foundations allows rats to burrow into buildings, so maintenance staff should keep masonry repaired and should block fan openings and other potential entrances.

It is easier to control rodents if they have no place to live. Outside equipment should be kept 23 to 30 cm (10–12 in) off the floor, shrubbery should be at least 10 m (30 ft) away from food facilities, and food-processing buildings are best surrounded by a grass-free strip 0.6 to 0.9 m (2–3 ft) wide covered with a layer of gravel or stones 2.5 to 3.8 cm (1–1.5 in) deep. This helps control weeds and rodents.

Eliminating rodent shelters. Crowded and unclean storage rooms provide sheltered areas where rodents can build nests and reproduce. Rodents love garbage and other refuse. Refuse is less attractive to rodents if it is stored 0.5 m (18 in) above the floor or ground. Storing waste containers on concrete blocks eliminates hiding places under the containers. Waste containers should be made of heavy-duty plastic or galvanized metal and have tight-fitting lids.

Foodstuffs should be stored on racks at least 15 cm (6 in) above the floor or away from the walls. A white strip painted around the edge of the floor in storage areas reminds workers to stack products away from the walls and helps show up rodent tracks, droppings, and hair.

Eliminating rodent food sources. Storing food and supplies properly and proper cleaning can help eliminate rodent food sources. Workers should clean up spills promptly, sweep floors regularly, and take out trash regularly. They should store food ingredients and supplies in tightly sealed containers.

Poisoning. Poisoning can be a good way to kill rodents, although the poison can also be dangerous to human beings. The most common rodenticides are fumarin, warfarin, and pival. (Warfarin is not used much in commercial establishments.) Rodents need to consume these poisons several times (multidose poisons) before they die, so adult humans who accidentally eat poisoned bait once are not in danger.

Multiple-dose poisons are safer than most other poisons, but workers still need to prepare and apply them according to the directions. Ideally, workers should put poison along rodent runways and near feeding sites. To make sure that humans do not accidentally eat the bait, workers should place it in bait boxes or under rodent shelters. Workers should put out fresh bait daily for at least 2 weeks to make sure that the poison works.

Rodenticides are commercially available in several forms. They are sold as ready-to-use baits to go in plastic or corrugated containers near rodent runways; in pellet form, mixed with grain to go in rodent burrows and dead spaces between walls; in small plastic packages to go in rodent hiding places; in bait blocks; and as salts to mix with water. The sanitarian or pest control operator should keep records of all bait containers so that they can be inspected and replaced easily. If rodents have not eaten the bait after two or more inspections, it should be moved.

If an operator needs to kill rodents immediately, he or she should use single-dose (acute) poisons such as red squill and zinc phosphide. Operators can mix these poisons with fresh bait, such as meat, cornmeal, or peanut butter. Operators should prepare and use these baits according to manufacturer's directions. Unfortunately, some of the single-dose poisons only work against Norway rats.

Sanitarians should place baits in several places, because rodents often travel only a short distance from their shelters. If there is enough food and shelter, rats stay within 50 m (150 ft) and mice within 10 m (30 ft) of their homes. If baits are too far apart or are in the wrong places, rodents may not find the poison. Several baits should be placed wherever there are signs of recent rodent activity. Because they often carry food to their nests, rodents may die there and workers may smell their bodies decomposing. Workers need to remove and burn or bury dead rodents.

Bait is one of the best ways to kill rodents, but rats that have been ill or in pain without dying after eating a poison may avoid the bait. They are also cautious if dead or dying rats are close to a bait. Sanitarians can overcome these problems using a nonpoisoned bait for about 1 week and then using the same bait with rodenticide. Prebaiting is very important with single-dose poisons but is not recommended with fumarin and warfarin. Prebaiting is not needed for mice, because their avoidance instincts are weaker.

Tracking powder. Toxic tracking powder kills rats, and nontoxic powder shows where rats are and how many there are. Toxic powders contain an anticoagulant (fumarin or warfarin) or a single-dose poison. The poison kills rodents when they groom themselves after running through the powder. Powders work well if the food supply is abundant and rodents are therefore less interested in bait. Tracking powders work better against mice than against rats.

Gassing. Operators should only use this technique if other eradication methods do not work. If necessary, a professional exterminator or a thoroughly trained employee should gas rodent burrows with a compound such as methyl bromide. They should not gas rodent burrows if they are less than 6 m (20 ft) from a building because the burrows may go under the building.

Trapping. This is a slow but generally safe method of killing rodents. Operators should place traps at right angles to rodent runways, with the baited or trigger end toward the wall. Any food that appeals to rodents works as bait. Operators should check traps daily, removing trapped rodents and replacing bait as necessary. Operators should only use traps with other methods of control and should use as many traps as possible. Rats are shy and adaptable and can avoid traps as well as bait. A glueboard, which traps a mouse's feet in glue, is a good mousetrap. The pest control operator throws away the disposable tray and trapped mouse and replaces it with a new tray.

Ultrasonic devices. Ultrasonic sound waves are supposed to keep rodents away. However, if rats are hungry, they ignore the sound. These are not effective rodent controls.

BIRDS

Birds such as pigeons, sparrows, and starlings can be a problem in food facilities. Their droppings are ugly and can carry microorganisms that cause disease in human beings. Birds can also carry insects into food establishments.

Proper management and sanitation can reduce the bird population. Birds are attracted by food waste, so proper waste disposal keeps them away. Screens on doors, windows, and ventilation openings help keep birds out.

Trapping is an acceptable way to control birds. Wires that give a mild electric shock and pastes that keep birds away also stop them from nesting near food establishments. But electric wires are expensive and need frequent inspection and maintenance. Flashing lights and noise have little effect on birds, because they soon get used to this equipment. Other techniques that work if they are kept up are removing bird nests and harassing birds by spraying them with water. The best way to get rid of birds is to use an exterminator who specializes in bird control. A professional exterminator has the proper expertise and equipment and can use the right chemicals safely.

Chemical poisons can reduce the number of birds, although these poisons are not safe inside a food establishment. Operators once used strychnine-coated cereal grains, but some local regulations now restrict use of strychnine. Operators should remove dead birds so that dogs and cats do not eat them and suffer from secondary poisoning. Another compound that controls the bird population is 4-aminopyridine. This compound kills birds and also makes the birds that eat it make distress sounds and behave abnormally, which frightens away other birds. Azacosterol makes birds temporarily sterile but is only approved for control of pigeons. This biological control method carries less risk than other compounds but only provides a long-term solution (in other words, it does not remove the problem immediately), especially in species like pigeons that have a long life. Temporary sterility is not a very valuable bird control for sanitarians who need to get rid of birds immediately.

Traps with live decoys are best for controlling birds. The biggest limitation of bird trapping is the cost of labor and materials.

USE OF PESTICIDES

Sanitarians should not spray insecticides in food areas while they are in operation; they should wait until after the shift, until the weekend, or other times when the food establishment is closed. They should be careful that the insecticide does not spatter or drift onto other surfaces or food. Insecticide dusts usually contain a dry form of the same toxic compounds in sprays. Dusts take more skill than sprays to apply. Only professional pest control operators should use them.

Before using insecticides, even those approved for edible food products or supply storage areas, workers should cover all exposed food and supply items or, better yet, remove them to another area. Sanitarians need to wash spraying equipment after using it by scrubbing with a cleaning compound and hot water and then rinsing. Sanitarians should not use products containing residual-type insecticides on any surfaces that food will touch. A fumigation procedure is not recommended unless other methods have not worked, and even then a professional fumigator should do it. Regular plant staff or supervisors should not try to fumigate the facility unless they have the right training. Even when professional fumigators are used, plant managers should make sure that they take all the right safety precautions

The following precautions are suggested by the National Restaurant Association (1992) when applying pesticides:

1. Make sure all pesticide containers have clear labels.
2. Check that exterminators have work insurance to protect the establishment, employees, and customers.
3. Follow instructions when using pesticides. Only use these chemicals for the purposes on the label. An insecticide that works against one type of insect may not work against other pests.
4. Use the weakest poison that will destroy the pests.
5. Use oil-based and water-based sprays in appropriate places. Use oil-based sprays where water could cause an electrical short circuit, shrink fabric, or cause mildew. Use water-based sprays in places where oil may cause fire, damage to rubber or asphalt, or a bad smell.
6. Avoid prolonged exposure to sprays. Wear protective clothing while applying pesticides, and wash hands afterwards.
7. Be careful not to contaminate food, equipment, and utensils with pesticides.
8. Call a doctor, poison control center, or fire department rescue squad if accidental poisoning occurs. If you cannot get immediate help, make the victim vomit by

putting a finger down the victim's throat. Then give the victim 2 tablespoons of Epsom salts or milk of magnesia in water, followed by one or more glasses of milk or water. If strychnine poisoning occurs, administer 1 tablespoon of salt in a glass of water within 10 minutes to induce vomiting, followed by 1 teaspoon of activated charcoal in half a glass of water. Lay the victim down, and keep the person warm.

Chemical pesticides cannot substitute for proper sanitation. Proper sanitary practices work better and are cheaper than pesticides. Even with good pesticides, pests keep coming back if conditions are not sanitary.

Food facilities can minimize contamination if they store only essential pesticides on the premises. Sanitarians should inspect pesticide supplies regularly to check the inventory amount and the condition of the chemical. Sanitarians should store pesticides carefully:

1. Store pesticides in a dry area at a temperature below 35°C (95°F).
2. Store pesticides in a locked cabinet away from food-handling and food storage areas. Store these compounds separately from other chemicals, such as cleaning compounds.
3. Do not transfer pesticides from their labeled package to any other storage container. Storage of pesticides in empty food containers can cause pesticide poisoning.
4. Place empty pesticide containers in plastic bins marked for disposal of hazardous wastes. Even empty containers can be dangerous because they still contain small amounts of toxins. Burn paper and cardboard, but do not burn empty aerosol cans. Follow local regulations about use and disposal of pesticides.

INTEGRATED PEST MANAGEMENT

Integrated pest management (IPM) uses several methods together to control pests. IPM aims to control pests inexpensively using environmentally sound techniques, including biological control. The goals of IPM are to use pesticides wisely and to find alternatives whenever possible. With IPM, operators can use less pesticide because other control methods are working at the same time.

The term IPM implies that pests are "managed" rather than removed. However, the final goal of pest management in food processing is to prevent or get rid of pests. Pest control practices include inspection, housekeeping, physical and mechanical methods, and chemical methods. Integrated use of these methods allows economical, effective, and safe pest management.

Inspection

Inspection is a preventive control measure. It is time consuming but important and cost effective in the long run. Inspection is a critical part of IPM. Inspection can find problems early on, pick up potential problems before they begin, and check that the sanitation program is working.

Sanitarians should carry out formal inspections at least monthly. The inspector should check thoroughly for signs of pests and evaluate how well the pest management program is working. If possible, well-qualified inspectors from outside the plant (e.g., corporate staff inspector, consultant, or contracting inspection service representative) should inspect the facility.

Plant personnel should carry out ongoing informal inspections in their own work areas. Supervisors should encourage employees to be aware of sanitation problems that make pest control less effective.

Inspectors should check raw materials, manufactured or prepared products, outside the building, inside facilities, and equipment. Inspectors should use a flashlight and have the correct tools to open equipment and sample containers. Sanitarians should put together an inspection form to make sure the inspector does not miss anything and to record results. The forms give a record of potential problems and how problems are solved.

Housekeeping

Employees need to understand and be responsible for maintaining the facility's standards of cleanliness and cleaning schedules. Cleaning must be continuous in many areas, because even small amounts of product residues can attract pests, and undisturbed spaces can provide homes for pests.

Physical and Mechanical Methods

Because many pesticides that used to be common can no longer be used for pest control, physical and mechanical methods are important. Examples are rodent traps, glue boards, and electric fly traps. These methods do not usually contaminate the area and can make an IPM program more effective overall. Use of very hot or cold temperatures, sometimes combined with forced air movement, can help control pests.

Insects need the right amount of moisture. Foods with a low moisture content (especially below 12%) do not usually support growth of insects.

Different forms of radiation (radio frequencies, microwaves, infrared and ultraviolet light, gamma rays, X rays, and accelerated electrons) can disinfect food products, but not all of these methods are effective and practical. Commercial food operations sometimes use gamma rays, X rays, and accelerated electrons to get rid of insects.

Chemical Methods

Pesticides and other chemicals such as repellents, pheromones, and sticky materials are also an important part of IPM. However, IPM uses fewer chemicals because this is not the only method used. Staff who apply pesticides must know how to use each chemical so that it is safe, legal, and effective.

The Environmental Protection Agency (EPA) classifies pesticides as general use and restricted use. Those classified as restricted use are more likely to damage the environment or hurt the person using them. Therefore, only certified applicators can purchase and use these pesticides. States offer an EPA-approved training and certification program for staff who will need to use these chemicals.

The pesticide storage area should be large enough to store normal supplies of pesticide materials neatly and safely. The pesticide storage area should be located away from food, in a separate building if possible. This storage area should have power ventilation exhausting to the outside and should never be cross-ventilated with food-processing or food container storage areas. This storage area should be totally enclosed by walls and locked, so that only authorized staff have access to pesticides. The storage area should be dry and at the right temperature to prevent the chemicals from breaking down or reacting. Pesticide containers should have a visible label, and the sanitarian should check the inventory regularly. Pesticide handling and application equipment should include rubber gloves, protective outer garments, and respirators such as dust masks or a self-contained breathing apparatus (SCUBA).

SUMMARY

- The pests that cause the most problems in the food industry include the German cockroach, American cockroach, Oriental cockroach, housefly, fruit fly, Norway rat, house mouse, pigeon, sparrow, and starling.
- The most important parts of pest control are preventing pests from getting into food establishments and getting rid of their shelter areas and food sources.
- If pests become established, pesticides, traps, and other controls are essential.
- Pesticides and traps can never replace good sanitation practices.
- Pesticides are toxic, so sanitarians should select, store, use, and dispose of them carefully. A trained employee can handle pesticides, but it is better to employ a professional exterminator for dangerous chemicals.

BIBLIOGRAPHY

Anon. 1976. *Plant Sanitation for the Meat Packing Industry.* Office of Continuing Education, University of Guelph and Meat Packers Council of Canada.

Hill, D. S. 1990. *Pests of Stored Products and Their Control.* CRC Press, Boca Raton, Fla.

Katsuyama, A. M., and Strachan, J. P. 1980. *Principles of Food Processing Sanitation.* The Food Processors Institute, Washington, D.C.

Marriott, N. G. 1994. *Principles of Food Sanitation,* 3d ed. Chapman & Hall, New York.

Mills, R., and Pedersen, J. 1990. *A Flour Mill Sanitation Manual,* p. 55. Eagan Press, St. Paul, Minn.

Restaurant Association Education Foundation. 1992. *Applied Foodservice Sanitation,* 4th ed. John Wiley & Sons, in cooperation with the Education Foundation of the Restaurant Association, New York.

Robinson, W. H. 1973. Insect pests of food service operations. In *Proceedings of the Conference on Sanitation and Food Safety,* pp. 69–73. Extension Division, State

Technical Services, Food Science and Technology Department and Virginia School Food Service Department, Blacksburg.

Shapton, D. A., and Shapton, N. E., eds. 1991. Buildings. In *Principles and Practices for the Safe Processing of Foods,* p. 37. Butterworth-Heinemann Ltd., Oxford.

Troller, J. A. 1993. *Sanitation in Food Processing,* 2d ed. Academic Press, New York.

STUDY QUESTIONS

1. Which insects cause most problems in food establishments?
2. How do pests spread disease?
3. How would you find out if a facility is infested with cockroaches?
4. How can you prevent cockroach infestations?
5. What are the best ways to control flies?
6. What is the difference between residual and nonresidual pesticides?
7. List four ways to control rodents.
8. List four precautions when using pesticides.
9. What is integrated pest management?

TO FIND OUT MORE ABOUT PEST CONTROL

1. Contact the National Pest Control Association, (1-800) 678-6722, and ask for information.
2. Contact a sanitation and pest control products and services company (e.g., Diversy Corporation, (1-313) 458-5000 or Ecolab Inc., (1-612) 293-2233), and ask for information on pest control.
3. Use the Yellow Pages to find local pest control services. Are there any who specialize in pest control in food establishments? Call and ask them about the problems they encounter most often in your area.
4. If you work in food production or foodservice, ask your employer what is done to control pests at your facility.

COCKROACHES AND BACTERIA

People are often afraid of cockroaches and consider them dirty. They also carry an unsanitary collection of bacteria, fungi, helminths, and viruses. To find out how likely cockroaches are to cause disease, the authors of this study collected cockroaches from several places to find out which bacteria they were carrying. They collected cockroaches from a hospital nursing area and out-patient area, a swimming pool pool-side and toilet area, low-income apartments, and commercial kitchens. The 157 samples (each sample contained 5 to 10 cockroaches) contained 56 types of bacteria, 14 of them known to cause disease in humans. Cockroaches collected from the swimming pool pool-side carried fewer bacteria, probably because of the chlorine in the water. The authors conclude that cockroaches can be extremely hazardous to human health and that pest management programs are essential, especially in commercial kitchens and hospitals.

Source: Rivault, C., Cloarec, A., and Le Guyader, A. 1993. Bacterial load of cockroaches in relation to urban environment. *Epidemiol. Infect.* 110:317.

CHAPTER 10

Quality Assurance and
Hazard Analysis Critical Control Points

ABOUT THIS CHAPTER

In this chapter you will learn:

1. What quality assurance (QA), total quality management (TQM), hazard analysis critical control points (HACCPs), and good manufacturing practices (GMPs) mean to the food industry
2. How management can support and design QA and HACCP programs
3. The steps involved in QA programs, including inspecting, sampling, testing, training, record keeping, problem solving, and product recalls, if needed
4. The seven steps to designing and using a HACCP program

INTRODUCTION

History

The food industry has used organized quality assurance (QA) programs since the late 1970s to make more acceptable food products. QA programs monitor the microbiology of raw ingredients used in food processing and the wholesomeness and safety of the finished food products. Today, QA and sanitation programs are even more vital for the food industry. Regulatory agencies have begun voluntary QA programs. Also, many professional or trade associations have developed manuals and other materials on QA. Food scientists who belong to these associations have been able to use these materials to begin or improve QA programs in their own organizations.

Why Use a QA Program?

To be safe and profitable, all food establishments need a QA program that stresses sanitation. The government has set hygienic standards for foods. The production department is responsible for meeting these standards. But it may be impractical for production staff to measure and monitor sanitation and still maintain high productivity and efficiency. Therefore, organizations use a QA program to monitor each phase of the operation (raw materials, critical processing points, sanitation, and finished products). The program should include every step of processing to help control the safety of all products. Staff should work as a team to meet sanitary standards and ensure safe food.

QA as an Investment

Lack of QA may be expensive. The cost of a QA program can be offset by a better image for the product, less likelihood of product liability suits, consumer satisfaction with a consistent and wholesome product, and improved sales.

QUALITY ASSURANCE (QA)

Definition of QA

Quality refers to how acceptable a product is to a consumer. Quality has components that can be measured and controlled. QA is the planned and systematic inspection, monitoring, evaluation, and record keeping required to maintain an acceptable product. QA techniques can be applied to sanitation as well as product manufacture.

QA in the Food Industry

The food industry uses QA principles to make sure that its sanitation practices are effective. If employees have not cleaned something adequately, they can take action to correct the problem. Larger operations often use a daily sanitation survey with checks and forms. Visual inspections must be more than a quick glance, because film buildup on equipment can hold microorganisms that cause spoilage and food poisoning.

Components of a QA Program

A QA program should include the following components:

1. Clear objectives and policies
2. Sanitation requirements for processes and products
3. An inspection system that includes procedures
4. Specified microbial, physical, and chemical values for products

5. Procedures and requirements for microbial, physical, and chemical testing
6. A personnel structure, including an organizational chart for the QA program
7. A QA budget to cover related expenses
8. A job description for all QA positions
9. An appropriate salary structure to attract and retain qualified QA staff
10. Constant supervision of the QA program and regular reports of the results

The QA Department

Large-volume processing plants should have a separate QA department. The QA department can measure components that production departments do not monitor because of time shortages and their emphasis on efficiency. QA staff respond to problems, interpret results in practical and meaningful ways, and assist with corrective actions. QA staff must cooperate with other departments to maintain the efficient production of acceptable products. Ideally, QA should be integrated into the strategy of the organization.

Responsibilities of a Sanitation QA Program

Before the program can be designed, QA staff need to identify the following factors:

1. Criteria for measuring what is acceptable (e.g., microbial levels)
2. Appropriate control checks at key production points
3. Sampling procedures (e.g., sampling times, numbers to be sampled, and measurements to be made)
4. Analysis methods

The responsibilities of a QA department include the following:

- Inspect the facility and equipment at least daily.
- Work with research and development (R&D) and production and marketing (sales) departments to develop specifications and standards.
- Develop and implement sampling and testing procedures.
- Implement a microbial testing and reporting program for raw products and manufactured products.
- Evaluate and monitor staff hygiene.
- Check that all aspects of the QA program meet national and local regulations, company guidelines and standards, and equipment manufacturers' recommendations.
- Inspect production areas for hygienic practices.
- Evaluate the performance of cleaning compounds, cleaning equipment, and sanitizers.
- Implement a system for handling waste products.

- Report testing and inspection results to each area of management so that corrective action can be taken, if needed.
- Analyze microbial levels in ingredients and the finished product.
- Educate and train plant staff about hygiene, sanitation, and QA.
- Work with regulatory officials to correct technical problems.
- Work with various departments to troubleshoot problems.
- Manage the QA department, including development and control of the budget and recruitment and training of QA staff.
- Maintain up-to-date records of all data and corrective actions.

The Role of Management in QA

The success or failure of a sanitation program depends on whether management supports the program. Managers may not be interested in QA because it is a long-term program and may seem to be expensive. Lower and middle management will have difficulty making a QA program work if top management does not fully understand the concept.

Some more progressive management teams are more enthusiastic about QA. They realize that a QA program can be a promotion tool, increase sales, and improve the image of their organization. All QA employees must understand the importance of their jobs. Management can make QA glamorous and exciting through a job enrichment program. For more information on job enrichment programs, consult a management text or management journals.

Structure of QA Programs

It is important to decide who is responsible for QA and how the program will be managed. In the most successful efforts, the QA program is run by top management, not just part of production. QA staff should report directly to top management, not production management, but the QA and production departments should work closely together. The QA department makes sure that sanitation problems are corrected and that the quality and stability or keeping quality of the final product are acceptable. Figure 10.1 shows the areas of responsibility of the administrator of the QA program.

Both large and small organizations need to develop a QA program. A smaller organization's program may be scaled down but should still stress hygienic methods and monitoring so that the final product is consistent and wholesome. Smaller organizations should assign a vice president, superintendent, or other high-level employee to manage the QA program.

The daily functions of a sanitation QA program should be delegated to a sanitarian. This employee must be allowed the time and resources to keep up on methods and materials needed to maintain sanitary conditions. Management should clearly define the responsibilities of the sanitarian using a written job description and the organization chart. The role and position of the sanitarian should be made clear to all

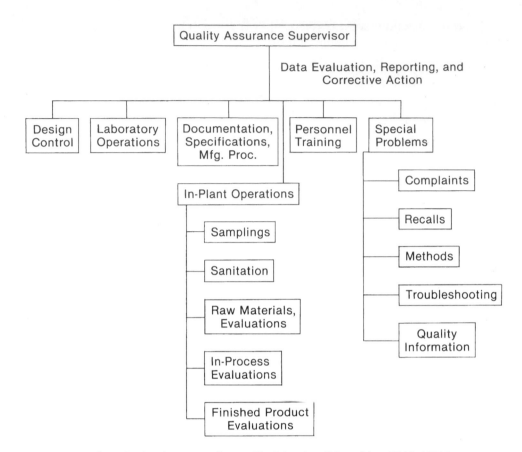

FIGURE 10.1. Organizational structure for specific QA tasks. *(Adapted from Webb, 1981.)*

staff. The sanitarian should report to the level of management with authority over general policy. This position should be equal to managers of production, engineering, purchasing, and other similar departments so that the sanitarian has the authority to administer an effective sanitation program. Although responsibilities may be combined in smaller-volume operations, management should still clearly define the responsibilities. The sanitarian should understand his or her responsibilities and how the position fits in the company structure. Figures 10.2 and 10.3 show examples of how the plant sanitarian could fit in the QA program of a large and a small processing operation.

A good QA program is run by one or more employees with technical training. The QA director or manger should have experience in food processing and/or preparation. Other QA staff can be less highly trained, provided they show interest, leadership, and initiative. Workshops, short courses, and seminars are often available to help train new workers.

Records and reports are important parts of a QA program. Records are useful in dealing with problems such as customer complaints, selecting suppliers, and cost accounting.

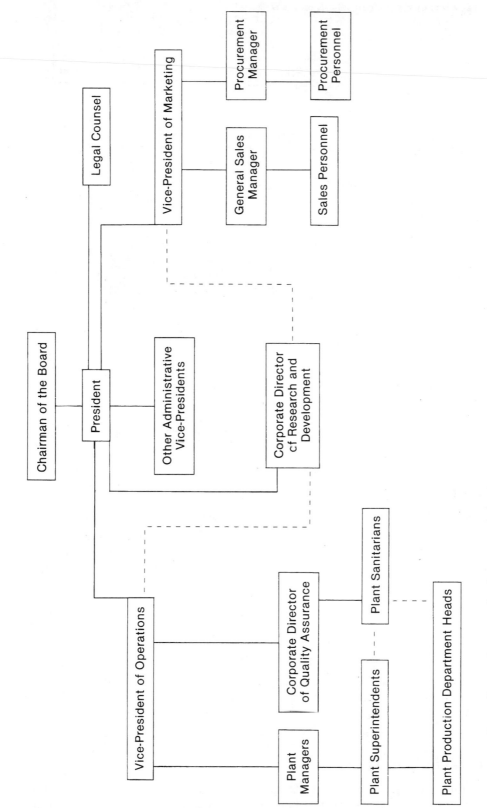

FIGURE 10.2. Chart reflecting the status of a plant sanitarian in a large organiza-

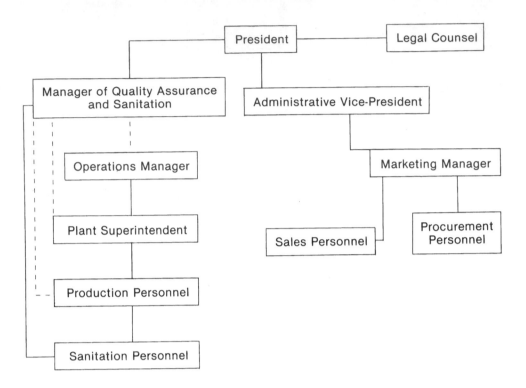

FIGURE 10.3. Chart reflecting the status of a plant sanitarian in a small organization.

Monitoring During Production

QA personnel should monitor sanitation at all stages of production, from receiving raw materials to packaging finished products. A good monitoring program can pick up changes in production that may affect the acceptability of the finished product. Monitoring also makes sure that product flows properly through the production cycle and detects hygiene or other production problems.

Introducing an Effective Quality Assurance Program

Management must be careful to prepare staff for the more unified system of control. To help prevent resistance and to encourage a new attitude, management should explain to staff why they are making the changes and how the program fits into the philosophy of the company.

Elements of a Total QA System

QA staff must outline the processes that occur in the operation. One way is to look at the physical layout of the plant. It is helpful to make a list of the rooms or areas and draw a flow diagram of the production processes, starting with the receiving area and ending with the shipping area for finished goods. Managers list the activities for each area, as well as how to control each activity (e.g., "examine raw materials," "clean equipment," or other sanitation checks). Managers then decide how often to test or check the activity and how to keep good records. For each production area, one person should be responsible for all of the controls or inspections, either an employee or an outside contractor. Once completed, management can convert the outline into a manual for the staff responsible for QA.

Sanitation Inspections

A total QA system should include a sanitation inspection of plant facilities and operations, including outside areas and storage areas on plant property. Figure 10.4 shows the daily sanitation report proposed by the U.S. Department of Agriculture (USDA). In a total QA system, a plant official should make the sanitation inspection and record the findings. If he or she discovers problems, staff should follow a plan for corrective action. Corrective action might include recleaning or closing off an area until a repair is completed. QA staff should inspect any point where the product could be contaminated (e.g., from container failure, moisture dripping, or grease escaping from machinery onto the food or surfaces that come into contact with the food) often and in an organized manner.

Training New Employees

New employees need to be familiar with all aspects of the operation. Managers can use the QA program to teach new employees about the production process and about good hygiene. Education should include basic information on food handling, cleanliness, and hygiene. An employee orientation checklist is helpful to make sure that trainers cover all topics. Managers should make sure that employees have ongoing training to remind them about good sanitation.

Evaluating QA Programs

QA staff use sensory and microbial testing to evaluate the sanitation QA program. Inspectors rely mostly on appearance to evaluate cleanliness. For most inspectors, a production area that looks clean, feels clean, and smells clean is satisfactory for production. But a good QA program must include concrete tests to evaluate microbial contamination.

Monitoring and Testing Procedures

The purpose of monitoring is to avoid problems with product safety and acceptability. The products and surfaces to be tested depend on the foods produced, production steps, and importance of the various surfaces for safety and acceptability of the food. The accuracy and precision, time requirements, and costs of the monitoring

U.S. Department of Agriculture
Animal and Plant Health Inspection Service
Meat and Poultry Inspection Program
DAILY SANITATION REPORT

Establishment Name	Est. No.	Date

INSTRUCTIONS: Prepare original and one copy signed by inspector and plant management. Inspector files original and gives copy to plant management.

Under the abbreviations "Pre-Op." (Observations made prior to the start of operations) and "Oper." (Observations made after operations have begun), record as appropriate the following codes. "N.O." (Not Observed), "A.C." (Acceptable), "Def." [Deficiency(s)].

GENERAL AREA	PRE-OP.	OPER.	REMARKS (Enter "General Area:" No., specific description of deficient areas, equipment, etc.)	ACTIONS TAKEN AND DOWNTIME (Enter "General Area" No.)
1. Ante-Mortem Areas				
2. Outside Premises				
3. Floors				
4. Walls				
5. Windows, Screens, etc.				
6. Ceilings and Overhead Structures				
7. Doors				
8. Rails and Shackles				
9. Equipment: a. Product Zone b. Nonproduct Zone				

10. Freezers and
    Coolers

11. Ice Facilities

12. Dry Storage Areas

13. Lights

14. Welfare Facilities

15. Employee:
    a. Dress
    b. Hygiene
    c. Work Habits

16. Handwashing and
    Sanitizing

17. Rodent and
    Insect Control

18. General
    Housekeeping
    a. Production area
    b. Nonproduction area

19. Production
    Practices

20. Other

RECEIVED BY ESTABLISHMENT OFFICIAL (*Signature*)   INSPECTOR(S) SIGNATURE   Page No.
                                                                                      ____ of ____

MP FORM 455   PREVIOUS EDITIONS OBSOLETE
Feb. 1974

FIGURE 10.4. Sample daily sanitation report. (*USDA, 1974.*)

159

program will vary for different operations. Data are more likely to be interpreted properly if QA staff use statistics to analyze the data. Management also need to understand the benefits and limitations of the tests used. For example, bacterial clumps from the contact method of sampling should give lower counts than the swab method, which breaks up cell clumps.

The monitoring program should include a way to evaluate the information from testing. Managers should set acceptable and unacceptable limits for test results for different testing times (such as after cleaning and sanitizing or during manufacture). Regular monitoring under the same conditions can show trends in contamination. The QA manager can use this information to set realistic guidelines for production.

Microbial testing of food contact surfaces can measure the effectiveness of a QA program. A monitoring program can also show potential problem areas in the operation and be used to train the sanitation crew, supervisors, and QA employees.

Recall of Unsatisfactory Products

Product recall means bringing back products from the distribution system (transportation, storage, or store shelves) because they are unsatisfactory in some way. Every food business could potentially have a problem requiring product recall. Management can preserve the public image of the business during a recall if it is well organized.

Product recall may be voluntary if the firm finds the problem, or involuntary if the FDA finds the problem. FDA has three classes of product recall, depending on the health hazard involved:

Class I: Use of or exposure to the product will probably cause a serious public health hazard, perhaps even death, e.g., if it is contaminated with something toxic (chemical or microbial).

Class II: Use of or exposure to the product may cause a temporary health hazard, but a serious public health hazard (death) is unlikely, e.g., if it is contaminated with food infection microorganisms.

Class III: Use of or exposure to the product will not cause a public health hazard, e.g., if it is wrongly labeled.

If the monitoring program shows that the plant has products that are unsafe, management should use a recall plan. The recall plan needs to:

1. Collect, analyze, and evaluate all information related to the product.
2. Decide how quickly the recall can take place.
3. Inform all company officials and regulatory officials.
4. Provide instructions for the staff who will implement the recall.
5. Make sure that no more of the affected product lots are shipped.
6. Issue news releases for consumers about the product, if needed.
7. Inform customers.
8. Inform distributors and help them to track down the product.

9. Return all products to one location where they can be isolated.
10. Keep a detailed log of recall events.
11. Find out the nature, extent, and causes of the problem to prevent it from happening again.
12. Provide progress reports for company and regulatory officials.
13. Check the amount of product returned to see if the recall has been effective.
14. Decide what to do with the recalled product.

Sampling for a QA Program

QA staff inspect or analyze samples. A sample is a part of anything that represents all of what it came from. A sample should be:

- *Statistically valid,* which means that workers select the sample randomly, so that any part of the lot it came from had an equal chance of being chosen.
- *Representative,* which means it includes a proportionate amount of each part of the product. In other words, for a product with a coating, samples should include the inside as well as the coating; for a homogeneous product, samples should be taken from the surface and from the middle.

In general, products from each lot or production period should be sampled. The sample should be identical to the whole lot from which it was taken. Collecting, labeling, and storage of enough samples for inspection and analysis are major concerns of the QA staff.

Sampling Procedures

The methods for collecting solid, semisolid, viscous, and liquid samples must be consistent. The following is a suggested sampling procedure:

1. Collect only representative samples.
2. Record the product temperature at the time of sampling, if applicable.
3. Keep samples at the correct temperature: nonperishable items and those normally held at room temperature may be held without refrigeration; perishable and refrigerated items should be kept at 0° to 4.5°C (24–32°F); and frozen or special samples should be held at −18°C (0°F) or below.
4. Protect the sample from contamination or damage. Some plastic containers cannot be labeled with a marking pen because the ink can seep into the contents.
5. Seal samples to protect them.
6. Send samples to the laboratory in the original unopened container whenever possible.
7. Homogeneous bulk products or products in containers too large to be transported to the laboratory should be mixed, and at least 100 g (3.5 oz) of the product should be transferred to a sterile container under aseptic conditions. Frozen products can be sampled using an electric drill and 2.5-cm (1-in) auger.

The logical place to start control is with the raw materials when they arrive at the plant. When suppliers know that customers check their products closely, they are more likely to send good materials. QA staff should take more frequent samples from products that tend to vary than from those that are uniform. The number of samples needed to evaluate sanitation at any stage in production also depends on how much the product varies. QA staff should take and pool a small percentage of each lot of incoming raw material. Sanitarians should sample the product at one or more stages of production between the raw and finished products. During packing, staff can sample the finished product.

Samples usually consist of three to five specimens that are representative of the whole. One suggestion is to sample the square root of the whole number of product units; i.e., if 10,000 units are produced, 100 should be sampled. Products should be sampled daily to monitor the process effectively. The product should be evaluated against its specifications to make sure it does not exceed the limits set for levels of ingredients, contaminants, microbes, temperature, etc. Management should determine minimum or maximum values for each of these levels for each product. If three consecutive samples exceed the limit, production should stop, and the equipment or area should be cleaned and sanitized.

Basic QA Tools

QA staff need a variety of equipment and supplies for sampling and evaluating products, depending on the operation. QA staff will need the following information in order to evaluate products:

1. Ingredient specifications (the type and amount of ingredients in the product)
2. Approved supplier list
3. Product specifications (including packaging and labeling)
4. Manufacturing procedures
5. Sampling program
6. Monitoring program (analyses, records, and reports)
7. Local and federal regulations
8. Cleaning and sanitizing program
9. Storage, shipping, and receiving program
10. Recall program

Rating Scales

Two types of scales are used to measure characteristics of products and samples:

1. *Exact.* For components that can be measured using numbers, e.g., colony-forming units (CFUs) of bacteria, percent water content, parts per million of cleaning chemicals.

2. *Subjective.* For components that cannot be measured exactly. An individual makes a sensory judgment, e.g., taste, feel, sight, smell. Two types of scales are used to give number values to the subjective tests.

Scale 1	Scale 2
7, excellent	4, extreme
6, very good	3, moderate
5, good	2, slight
4, average	1, none
3, fair	
2, poor	
1, very poor	

TOTAL QUALITY MANAGEMENT (TQM) AND HAZARD ANALYSIS CRITICAL CONTROL POINTS (HACCP)

Total Quality Management

A sanitation program is one aspect of total quality management (TQM). TQM means a "right first time" approach to all aspects of the operation. The most important aspect of TQM is food safety and sanitation. Chapter 19 gives more information on TQM. For TQM to be successful, management and production workers must be motivated to improve the acceptability of the product. Everyone involved must be trained to understand TQM and their role in its success. Computer software is available for training , implementing, and monitoring TQM programs.

Hazard Analysis Critical Control Points (HACCP)

In the late 1960s, the National Aeronautics and Space administration (NASA) and the U.S. Army Natick Laboratories developed the HACCP concept. In 1971, the Pillsbury Company worked with these two organizations to adapt this rational approach to process control for the food industry. HACCP is a simple but very specific way to find hazards and make sure that the appropriate controls are in place to prevent hazards. HACCP is designed to prevent, rather than detect, food hazards. The FDA and FSIS has determined that HACCP can help prevent food safety hazards during meat and poultry production. Several scientific groups have recommended using HACCP, including the National Academy of Sciences (NAS) Committee on the Scientific Basis of the Nation's Meat and Poultry Inspection Program and the NAS Subcommittee on Microbiological Criteria of the Committee on Food Protection. These two committees know that HACCP makes sense for controlling the most critical areas of food production to make sure that food is safe and wholesome.

The HACCP technique looks at the flow of food through the process. It provides a way to determine which points are critical for the control of foodborne disease hazards and monitors each operation frequently.

- A *hazard* is something that could harm the consumer.
- A *critical control point (CCP)* is an operation or step where something can be done to prevent or control a hazard.

The HACCP concept has not yet become part of the food laws but is a valuable part of voluntary control of microbial hazards. This approach has become a trend-setter in food sanitation and inspection. Government regulators have approved the concept, and progressive food companies are adopting it.

HACCP PRINCIPLES

The HACCP concept has two parts:

1. *Hazard analysis* depends on an understanding of which chemical and physical contaminants and microorganisms can be present and the factors that affect their growth and survival. The factors that most often affect food safety and acceptability are contaminated raw food or ingredients, poor temperature control during processing and storage (time-temperature abuse), improper cooling (failure to cool to refrigerated temperature within 2 to 4 hours), improper handling after processing, cross-contamination (between products or between raw and processed foods), poor cleaning of equipment, failure to separate raw and cooked products, and poor employee hygiene and sanitation. During hazard analysis, sanitarians draw diagrams of the sequence or the manufacturing and distribution process to show points where microbes could contaminate the food, survive, and grow to levels that could cause foodborne illness.

2. *Critical control points* are the points where controls are needed to make sure that microbes do not reach the food, survive, or grow to dangerous levels. Not all steps in a process are critical, and it is important to separate critical and noncritical points. Sanitarians should look at each step and decide how important it is to the safety and acceptability of the food product. The flow diagram developed during hazard analysis can help to find critical control points. Monitoring may take the form of observation, physical measurements (temperature or pH), or microbial analysis. Microbial testing has often been too time consuming in the past, but several faster testing methods are now available. If the microbial status of the raw material is a CCP, microbial testing is appropriate. Microbial testing can have direct and indirect uses. A direct use of microbial testing is checking for microbes in raw materials, during processing, and in the finished product. An indirect use is monitoring the effectiveness of control points for cooking and cooling and for cleaning and employee hygiene. Sanitarians set critical limits (acceptable levels) for each monitoring procedure. Monitoring the critical control points and correcting any problems allow the food industry to protect consumers effectively and efficiently.

For HACCP to function well in the food industry:

1. Food processors and regulators must learn about HACCP.

2. Plant personnel must apply their technical skills to the program.
3. Sanitarians should drop HACCP control points that are not hazardous from the program.

DEVELOPING AND IMPLEMENTING HACCP

HACCP is an organized way to control food production and ensure safe food. Before implementing HACCP, processors must identify good manufacturing practices (GMPs) to make sure the plant produces safe food products. Each operation must have written sanitary standard operating procedures (SOPs). Operators must follow these SOPs to make sure the food is wholesome. Operators must then develop a step-by-step sequence of events in production so that they can identify potential hazards and critical control points. A HACCP program has seven basic steps.

Step 1. Conduct a hazard analysis. List the steps in the process where hazards can occur, decide how important each hazard is, and design ways to prevent the hazard. Sanitarians look at the risk from hazardous microorganisms or their toxins in a specific food and its ingredients. Hazard analysis can guide the safe design of a food product and show the CCPs that remove or control hazardous microorganisms or their toxins at any point during production. Hazard assessment is a two-part process. First, six potential hazards are characterized for the food. Second, this characterization is used to give the food a risk category.

Foods are placed into risk categories based on how many of six hazard characteristics the food has. Figure 10.5 shows a decision tree that can be used to decide whether the product has a potential hazard at any step during processing. If a product comes under hazard class A, it is automatically in risk category VI. The six hazards are as follows (hazards can also be chemical or physical):

Hazard A: A special class of nonsterile products intended for at-risk populations, e.g., infants, the elderly, people who are ill, or immunocompromized people

Hazard B: Products that contain ingredients that could be contaminated with microbes

Hazard C: Products manufactured by a process that does not have a controlled processing step that destroys harmful microorganisms

Hazard D: Foods that could be recontaminated after processing and before packaging

Hazard E: Foods that could be handled abusively and may become harmful during distribution or by consumers

Hazard F: Foods that are not heated after packaging or before they are eaten

Risk categories are based on these hazard characteristics:

Category O: No hazard

Category I: Food products that have one of the general hazard characteristics

Category II: Food products that have two of the general hazard characteristics

HACCP: Principles and Application

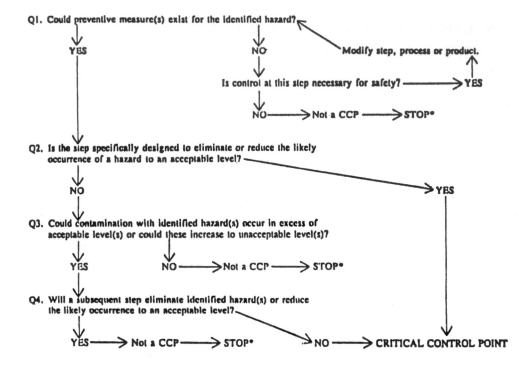

FIGURE 10.5. CCP decision tree. [Source: Pierson and Corlett (1992).]

Category III: Food products that have three of the general hazard characteristics

Category IV: Food products that have four of the general hazard characteristics

Category V: Food products that have all five of the general hazard characteristics: hazard classes B, C, D, E, and F

Category VI: Nonsterile food products intended to be used by at-risk populations, e.g., infants, the elderly, people who are ill, or immunocompromised people; sanitarians must consider all of the hazard characteristics

Step 2. Decide on CCPs that will control the hazards. Action must be taken to control the hazards at some point in food production, between growing and harvesting raw materials and eating the prepared food. Any point in food production where hazardous microorganisms should be destroyed or controlled can be a critical control point. An example of a CCP is heating for a certain amount of time at a certain temperature to destroy a specific microbial pathogen. Another temperature-related CCP is refrigerating the food to prevent hazardous organisms from growing, or adjusting the pH of a food to prevent toxins from forming. It is important not to confuse control points with critical control points that affect safety. A control point is any point during food production where loss of control does not lead to a health risk. Figure 10.5 can be used to decide whether a step is a control point or a CCP.

Step 3. Set target levels and tolerances (critical limits) that must be met to be certain that each CCP is under control. Critical limits are very important for safe control of a CCP. An example is cooking a meat product enough to eliminate the most heat-resistant microorganism that could be in the product. The critical limits (minimum and or maximum amounts) must be specified for time, temperature, and thickness of the product. To set critical limits, sanitarians need to know the maximum number of microorganisms that may need to be destroyed.

Step 4. Set up procedures to monitor CCPs. Monitoring is important to make sure that a CCP is within its limits. The person who does the testing or observing must keep records of the results. Failure to control a CCP is a critical defect and may cause the product to be hazardous or unsafe to eat. Monitoring is very important because critical defects can have serious consequences.

Monitoring should be continuous, if possible. Recorders can monitor pH, temperature, and humidity continuously. When monitoring cannot be continuous, the critical limit must be checked often enough to show that the hazard is under control. A statistically designed data collection program or sampling system will help make sure that testing is adequate.

The tests used to check CCPs must give results quickly. Therefore, physical and chemical measurements are more useful than slow microbial tests. Microbial testing is rarely useful in HACCP, but it can be used for random checks of how well CCPs are controlled. The person who does the testing and someone with management responsibility for sanitation should sign all results of monitoring.

Step 5. Decide what should be done to correct the problem (corrective action) when monitoring shows that a critical limit has not been met. The corrective actions must bring the CCP under control. The HACCP plan should include procedures for correcting critical defects and coordinating recalls. If a critical defect occurs, the production staff should place the product on hold until they correct the problem or test the product.

Step 6. Use good record keeping to show that the HACCP plan works. The HACCP plan and its results must be on file at the food establishment to document CCPs and action taken to correct critical deviations.

Step 7. Set up ways to prove that the HACCP plan is working properly (verification). Results of microbial, sensory, and chemical tests can prove that the hazards of the system were all identified when the HACCP plan was developed. Tests can also show that staff are carrying out the HACCP plan as written. This process is known as verification and may include:

1. Scientific or technical analysis to show that critical limits of CCPs are acceptable
2. Appropriate verification schedules, sample collection, and analysis
3. Documented, regular on-site review of all flow diagrams and CCPs in the HACCP plan to make sure they are accurate
4. Government regulatory actions to make sure that the HACCP plan is working

Tables 10.1 and 10.2 give simplified illustrations of HACCP for roasting and thawing chicken.

TABLE 10.1. Simplified Illustration of HAACP for Roasting Chicken

Process	Hazard	Critical Control Point	Standard	Monitoring Method	Action to Take If Standard Not Met
Purchase chicken	Natural bacterial contamination	None by consumer			
Cold storage	Natural bacterial contamination	Refrigerator temperature	Refrigerator air temperature at 4°C (<40°F)	Measure refrigerator air temperature	Lower thermostat setting
		Length of storage time (refrigerator)	2 days	Observation	Discard
Remove chicken from storage	Natural bacterial contamination	Personal hygiene	Thoroughly wash hands after contact with raw chicken	Observation	Rewash
		Sanitation of contact surfaces, including plates, utensils, sinks, and countertops	Wash, rinse, and disinfect after contact with raw chicken	Observation	Rewash and sanitize
Roast chicken	Natural bacterial contamination	Oven temperature	Oven temperature (conventional) at 163–177°C (325–350°F)	Measure oven air temperature	Reset thermostat
	Contamination of chicken surfaces by food handlers	Temperature of cooked chicken	Internal temperature at 82–85°C (180–185°F)	Measure internal temperature	Continue cooking until reaches appropriate temperature; if oven-cooked, remove from oven
	Cooking process may not be sufficient to kill bacteria				

Source: ©1991, The American Dietetic Association. *Food Safety for Professionals: A Reference and Study Guide.* Used by permission.

TABLE 10.2. Simplified Illustration of HACCP for Thawing Chicken Before Roasting

Process	Hazard	Critical Control Point	Standard	Monitoring Method	Action to Take If Standard Not Met
Freezer storage	Contaminating bacteria will survive	Freezer temperature	Freezer temperature −18°C (<0°F)	Measure freezer air temperature	Reset thermostat
Thawing	Insufficient thawing may lead to insufficient cooking, pathogen survival	Thawing time-temperature	Allow 1 day in refrigerator to thaw 2 kg (5-lb) chicken; check for microwave standards	Observation	Continue until thawed
	Bacteria can grow if portions are warm enough	Thawing time-temperature	Thaw in refrigerator	Measure refrigerator temperature	Lower thermostat
			Thaw in microwave oven	Follow appliance instructions	
			Thaw under cool running water	Observation	Reset water flow
	Drip from thawing chicken can contaminate surfaces and other foods	Sanitation of contact surfaces, including plates, utensils, sinks, countertops	Wash, rinse, and disinfect after contact with chicken or drip	Observation	Rewash and sanitize
Remove chicken from thawing surface and put on pan	Natural bacterial contamination	Personal hygiene	Thoroughly wash hands after contact with raw chicken	Observation	Rewash
		Sanitization of contact surfaces, including plates, utensils, sinks, countertops	Wash, rinse, and disinfect after contact with raw chicken	Observation	Rewash and sanitize
Roast chicken (see Table 10.1)					

Source: ©1991, The American Dietetic Association. *Food Safety for Professionals: A Reference and Study Guide.* Used by permission.

Success with HACCP

Employees need special training to use HACCP. Because of employee turnover, training must be continuous. If all employees understand the importance of keeping CCPs within the critical limits, cost-effective control can replace costly crisis management and outbreaks of foodborne illness can be reduced.

An effective HACCP program requires four steps:

1. *Management education.* Quality assurance personnel and higher management must understand the HACCP concept so that all staff will be committed to the program. Plant management create awareness of the program and should set a good example to other staff.
2. *Operational steps.* Management may need to change the plant design and operating procedures to make them more hygienic. Staff who carry out critical steps should be properly trained and experienced.
3. *Employee motivation.* Good working conditions can motivate staff to carry out HACCP. Supervisors may need to redesign tasks so that employees feel a sense of personal responsibility for the quality and safety of food products.
4. *Employee involvement.* Workers who are involved in problem solving are more likely to be committed to the program. Management should guide, not administer, the HACCP program and should consider recommendations from employee consultation groups or quality circles. All levels of employees must have long-term commitment to the program.

HACCP is important and may eventually become law, but can be difficult to start up. HACCP for one product may be different than for another product, and regulators need to know that the program is used and monitored on a day-to-day basis.

SUMMARY

- *Quality* refers to how acceptable a product is to a consumer. Quality Assurance (QA) is the planned and systematic inspection, monitoring, evaluation, and record keeping required to maintain an acceptable product.
- The QA department should be separate from the production department because of conflicting priorities. QA staff must cooperate with production staff and other departments to maintain the efficient production of acceptable products.
- The success or failure of a sanitation program depends on whether management supports the program.
- QA personnel should monitor sanitation at all stages of production, from receiving raw materials to packaging finished products.
- A total QA system should include a sanitation inspection of plant facilities and operations, including outside areas and storage areas on plant property.
- Product recalls may be voluntary or involuntary. FDA has three levels of product recall, depending on the risk to consumers.

- Samples used for testing should be statistically valid and representative.
- TQM means a "right first time" approach to all aspects of the operation. The most important aspect of TQM is food safety and sanitation.
- HACCP has two parts: (1) Hazard analysis looks at the production process to see where food safety problems could happen. (2) Critical control points (CCPs) are operations or steps where something can be done to prevent or control a hazard.
- Six categories are used to define risks for food products.
- HACCP programs are made up of seven steps.

BIBLIOGRAPHY

Baldock, J. D. 1979. Quality assurance for effective sanitation. In *Sanitation Notebook for the Seafood Industry,* p. III-59, G.J. Flick, Jr., et al., eds. Department of Food Science and Technology, Virginia Polytechnic Institute and State University, Blacksburg.

Bauman, H. E. 1987. The hazard analysis critical control point concept. In *Food Protection Technology,* p. 175. Lewis Publishers, Chelsea, Mich.

Carpenter, J. A. 1970. Planning your quality control system. *National Provisioner* 162(6):24.

Faig, H. L. 1979. Sampling. In *Sanitation Notebook for the Seafood Industry,* p. II-61. Department of Food Science and Technology, Virginia Polytechnic Institute and State University, Blacksburg.

Guzewich, J. J. 1987. Practical procedures for using the hazard analysis critical control point (HACCP) approach in food service establishments by industry and regulatory agencies. In *Food Protection Technology,* p. 91, C. W. Felix, ed. Chelsea, Mich.

Marriott, N. G. 1994. *Principles of Food Sanitation,* 3d ed. Chapman & Hall, New York.

Marriott, N. G., Boling, J. W., Bishop, J. R., and Hackney, C. R. 1991. *Quality Assurance Manual for the Food Industry.* Virginia Cooperative Extension.

National Advisory Committee on Microbiological Criteria for Foods. 1992. *Hazard Analysis and Critical Control Point System.* United States Department of Agriculture, Food Safety and Inspection Service, Washington, D.C.

Nickelson, R. II. 1979. Food contact surfaces indices of sanitation. In *Sanitation Notebook for the Seafood Industry,* p. II-51, G.J. Flick, Jr., et al., eds. Department of Food Science and Technology, Virginia Polytechnic Institute and State University, Blacksburg.

Pedraja, R. R. 1979. How to develop an effective quality assurance and sanitation program. In *Sanitation Notebook for the Seafood Industry,* p. III-45, G.J. Flick, Jr., et al., eds. Department of Food Science and Technology, Virginia Polytechnic Institute and State University, Blacksburg.

USDA. 1974. Sample daily sanitation report. U.S. Department of Agriculture, Washington, D.C.

STUDY QUESTIONS

1. Why do food companies need a QA program?
2. List five components of a QA program.
3. Describe the responsibilities of the QA staff.
4. What are the two types of rating scales used to measure products and samples? Give two examples of what could be measured with each scale.
5. What is the difference between a control point and a critical control point?
6. Why is microbial testing rarely used to check CCPs? When is microbial testing used more often?

TO FIND OUT MORE ABOUT QA AND HACCP

1. Apply HACCP to a recipe you would use in foodservice or at home. Use the seven steps to identify hazards, critical control points, target levels and tolerances, ways to monitor CCPs, how to correct problems, how to keep records, and how you could make sure that HACCP is working.
2. If your school or workplace has a cafeteria, ask the manager about the cafeteria's quality assurance program.
3. Look through issues of *FDA Consumer* magazine at the library for information on recent product recalls. List potential CCPs that might have prevented the recalls.

CASE STUDY: USE OF HACCP BY IN-FLIGHT CATERING SERVICES

Foodborne illness can affect both passengers and crews on commercial flights. It can be an air safety problem and is the leading cause of illness in pilots. In 1991, an outbreak of salmonellosis affected 415 passengers. After this outbreak, a group of researchers worked with the flight kitchen on a Greek island to introduce a HACCP program. The researchers used flow diagrams to show the steps used to prepare hot foods, cold foods, and desserts, and identified CCPs for each step:

Deliveries. Staff checked all raw supplies for sanitary delivery, temperature, date, time, quality, weight, and grades ordered.

Storage. Staff stored frozen foods at −18°C (0°F), chilled foods at 4°C (40°F), fresh vegetables between 8°C and 10°C (46°F and 50°F), and dry foods in dry stores. They separated raw and cooked food completely, recorded the temperatures of each storage area every day, and rotated the stock so that they used the oldest supplies first.

Preprocessing. Store prepared raw meats and vegetables in separate areas to prevent cross-contamination. They used different, color-coded knives, chopping boards, containers, and other equipment for raw and cooked foods, and defrosted meat and poultry in a refrigerator for 24 to 48 hours before cooking.

Cooking. Workers checked cooking temperatures with probe thermometers to make sure that the center of the food reached 70°C (158°F) for at least 2 minutes.

Blast chilling. Foodservice workers transferred the food to gastronome trays and began blast chilling within 30 minutes of the end of cooking. The temperature at the center fell to 3°C (37°F) within 90 minutes.

Chilled storage. Staff stored food at 4°C (40°F) and plated it the same or next day.

Plating. Food handlers plated food in an air-conditioned tray-setting area, wearing plastic gloves. They labeled the foods, blast-chilled them to −36°C (−33°F), and stored them at −25°C (−15°C) for up to 3 months.

Assembling and packing trays. Staff kept perishable food at 4 to 6°C (40–42°F) for less than 20 hours. Workers filled the aircraft trolleys and stored them at 4°C (40°F) with the doors open so that chilled air could circulate.

Distribution. Drivers used refrigerated vans to carry food to the aircraft to make sure the temperature of the food was not above 10°C (50°F) when delivered.

Cleaning. The cleaning schedule specified how and when to clean and disinfect the kitchen and equipment and which chemicals to use. Workers stored detergents, chemicals, and cleaning equipment in a separate room. They collected samples from food contact surfaces to make sure that cleaning was effective.

Training. The researchers trained the staff about food safety, cleaning procedures, handwashing, and proper control of intestinal and wound infections using videos and posters.

The HACCP program greatly improved the level of bacteria in the foods and the number of food samples that contained pathogenic microorganisms. It also improved the cleanliness of the food contact surfaces. Two factors helped to make the program successful:

- Workers and management showed great support for the program.
- Food handlers received the education they needed to understand food hygiene.

Source: Lambiri, M., Mavridou, A., and Papadakis, J.A. 1995. The application of hazard analysis critical control point (HACCP) in a flight catering establishment improved the bacteriological quality of meals. *J. R. Soc. Health* 115(1):26.

CHAPTER 11

Dairy Processing and Product Sanitation

ABOUT THIS CHAPTER

In this chapter you will learn:

1. How the design of dairy-processing plants affects the wholesomeness of dairy products
2. Which pathogens cause most problems in the dairy industry
3. About the types of soil in dairy plants
4. About cleaning and sanitizing compounds and equipment for dairy-processing plants

INTRODUCTION

The dairy industry is known for its hygienic designs and practices and high sanitation standards. This industry needs especially good sanitation practices to ensure that refrigerated dairy products are safe and keep well.

The physical and chemical makeup of dairy products, especially liquids such as milk, makes it possible to use automated cleaning for the processing facilities. Some of the features of automated cleaning include:

- Permanent piping with welded construction that allows automated cleaning
- Computerized control systems to organize complex cleaning sequences
- Automated daily cleaning-in-place (CIP) systems to keep tanks, valves, and pipes consistently clean
- Replacement of plug-type valves that require manual cleaning with air-operated valves to allow remote or automatic control of CIP solutions

- New designs of silo-type storage tanks, dome-top processors, homogenizers, plate heat exchangers, certain fillers, and the self-desludging centrifugal machine allow that CIP

CIP works best when sanitarians design and install a complete automated cleaning and sanitizing system. If properly designed, CIP can clean effectively and reduce labor costs.

The source of the milk supply has an enormous effect on the finished, pasteurized product. Even the best pasteurization cannot make up for poor quality and other problems created by undesirable bacteria in the raw-milk supply. Although pasteurization works well against pathogenic and spoilage microorganisms, it is not foolproof and cannot make up for an unsanitary raw supply or poor sanitation.

Polyphosphates, synthetic cleaning agents, and new cleaning and sanitizing equipment have changed cleaning operations. Sanitarians have designed specific cleaning compounds for different water conditions, types of metals, and types of soil that result in less corrosion. Cleaning compounds and sanitizing agents work together better to make both steps more effective.

Major Pathogens

Pasteurized milk caused a large outbreak of salmonellosis in 1985; ice cream has caused major outbreaks of staphylococcal food poisoning; and dairy products may have been responsible for sporadic cases of campylobacteriosis. But the pathogen that can cause the most severe problem in dairy products is *Listeria monocytogenes*.

Cheese contaminated with *L. monocytogenes* has caused several deaths as well as expensive product recalls. Many cattle carry *L. monocytogenes* in their intestines, and about 5 to 10% of milk contains this pathogen. *L. monocytogenes* grows at refrigeration temperatures, so even slightly soiled dairy products can become heavily contaminated during shipping and storage. Therefore, many dairy-processing plants have upgraded their sanitation standards to make sure that dairy products are not contaminated with *L. monocytogenes*.

Plant Construction

Location

A dairy plant should be as close as possible to the milk supply but should also be convenient for the population it will serve. A good supply of drinking water and good drainage and waste disposal are essential. Easy access to main highways allows prompt delivery of raw materials and supplies to the plant and easy shipment of finished products from the plant. The plant should also be close to an energy source that can handle its power demands.

Drainage and waste disposal are very important for good plant sanitation. Storm and sanitary sewers must be adequate. In rural areas and towns with small treatment facilities, dairy processors often have to provide their own waste-disposal facilities.

Also, the dairy plant should be near a good water supply that can meet the maximum demand of the plant even when it is at minimum pressure. The plant's peak water demand should not coincide with other peak demands on the city water supply. If the temperature or cost of city water is too high, the plant may need its own water source, such as a well or reservoir.

Floor Plan and Type of Building

Various agencies regulate the layout, construction, equipment, and utensils of dairy plants. Processors need to make sure that they follow local regulations.

Ventilation is important, especially in areas where heat is produced during processing. Ventilation should be appropriate for the room and should be flexible in case production changes. Processors may need to filter incoming air, especially if the plant is in a heavy industrial area. Also, processors may also need to control humidity, condensation, dust, and spores in the air.

Construction Guidelines

If construction is not planned carefully, the structure and equipment can contribute to contamination of food products. Planners should minimize the amount of overhead equipment so that food is less likely to be contaminated during maintenance. Overhead equipment is also difficult to clean. It is best to have a separate service floor to hold most of the ducts, pipes, compressors, and other equipment. This arrangement leaves a clear ceiling, which is easier to keep clean.

Other design and construction tips for good sanitation are:

- Treat all metal construction so that it does not corrode.
- Choose pipe insulation that resists damage and corrosion and endures frequent cleaning.
- Protect points where condensation may collect by installing a drainage collection system.
- Cover all openings with air or mesh screens and tight-fitting windows.

Designers should choose structure finishes that need very little maintenance. Walls, floors, and ceilings should not be porous. Floors should not be affected by milk, milk acids, grease, cleaning compounds, steam, and impact damage; epoxy, tile, and brick are good choices. Plants should not have painted floors unless there is no alternative. If a plant has paint on the floor or walls, it should be an acceptable grade for food plants. Floor drains should control insect infestation and odors. A slope of about 2.1 cm/m (0.25 in/ft) reduces accumulation of water and waste on the floor and helps control growth of *Listeria monocytogenes.*

Floor drains and ventilation systems can add to contamination instead of acting as a sanitation barrier. But a properly designed ventilation system with air filters can improve the air quality. Inexpensive filters can remove dust and other contaminants before drawing air into rooms.

Staff should clean floors, walls, and ceilings frequently and thoroughly. In the past, equipment layout was based on operational efficiency, and sanitation was a secondary factor. Now, sanitation is an equal consideration in the design process. Equipment must be far enough away from walls or partitions, and all surfaces must be easy to clean. All equipment should be accessible, cleanable, and designed for draining and sanitizing.

TYPES OF SOIL

In the dairy industry, soil is mainly minerals, lipids, carbohydrates, proteins, and water. Soil may also contain dust, lubricants, microorganisms, cleaning compounds, and sanitizers.

White or grayish films that form on dairy equipment are usually milkstone and waterstone. These films build up slowly on unheated surfaces if they are not properly cleaned or if the water is hard. Calcium and magnesium salts settle out when sodium carbonates are added to hard water. During cleaning, some of these may stick to equipment, leaving a film of waterstone. When heat denatures proteins (changes their shape), they cling to surfaces, and other salts stick to them, so that milkstone forms quickly on heated surfaces. Heated and unheated surfaces usually carry different types of soil, so each surface requires a different cleaning procedure. Milkstone is a porous deposit and hides microbes from sanitizers. It must be removed, usually with an acid cleaner to dissolve the alkaline minerals and remove the film. Heavy soil deposits require a strong cleaning compound. Fresh soil on an unheated surface dissolves more easily than the same soil that is dried or baked onto a heated surface.

The following principles can help reduce and remove soil deposits.

- Cool the surfaces before and immediately after emptying heated processing vats and other equipment.
- Rinse away foam and other residues after the production shift and before they dry.
- Keep soil deposits moist until cleaning starts.
- Rinse with warm (not hot) water.

More soil deposits form in ultra-high-temperature heaters if the milk contains air or is highly acidic, especially if the milk moves slowly or is not well agitated. Preheating and holding at a high temperature will reduce film deposits.

The surface affects how hard it is to remove soil. Pits in corroded surfaces, cracks in rubber parts, and crevices in poorly polished surfaces protect soil and microorganisms from cleaning compounds and sanitizers. The type of soil determines the cleaning method and concentration and type of cleaning compound.

SANITATION MANAGEMENT

Cleaned and sanitized equipment and buildings are essential for the production, processing, and distribution of wholesome dairy products. The largest part of the total cleaning cost is labor. Therefore, it is important to use the right cleaning compounds and equipment so that workers can clean the facility in a shorter period of time and with less labor.

The sanitarian should know how long it takes to clean each part of the operation with the equipment and cleaning compounds available. Cleaning tasks should be assigned to specific employees, who should be responsible for the equipment and area under their care. The sanitarian should give these assignments to workers in writing or post the cleaning schedule or assignments on a bulletin board. The cleansing operation should be monitored by sanitarian or another specific person.

Water

Most plant water supplies are not ideal, so sanitarians should choose cleaning compounds that will work well with the water supply or treat the water supply to allow the cleaning compound to work better. Suspended particles and hard water leave deposits on clean surfaces. Sanitarians can treat the water to remove suspended matter and soluble manganese and iron. Small amounts of water hardness can be overcome by sequestering agents in the cleaning compound. If the water is hard or very hard, it is usually cheaper to soften it before it is used for cleaning.

Cleaning Compounds

Like all cleaning compounds, those used in dairy plants are complex mixtures of chemicals combined for a specific desired purpose. Use of cleaning compounds in dairy-processing plants includes the following steps:

1. Prerinse to remove as much soil as possible and make the cleaning compound more effective.
2. Apply the cleaning compound to the soil to wet and penetrate it.
3. Use cleaning compounds to displace solid and liquid soils from the surface by making fat into soaps, breaking down proteins, and dissolving minerals.
4. Use dispersion, deflocculation, and emulsification to mix soil deposits in the cleaning solution.
5. Rinse well to prevent the dispersed soil from redepositing onto the clean surface.

The value of a cleaning compound is measured by the cost of cleaning a given area. Expensive cleaning compounds are often the most economical, because they save labor and energy and only a small amount of the cleaning compound is needed. See Chapter 5 for more information on cleaning compounds.

Table 11.1 shows the best cleaning compound, method, and equipment for cleaning dairy-processing plants and equipment. Workers should not use strong acids and alkalis for hand cleaning because they irritate the skin. Instead, they should use external energy, such as heat and force, with milder cleaning compounds.

TABLE 11.1. Optimal Cleaning Guidelines for Dairy-Processing Equipment

Cleaning Application	Cleaning Compound	Cleaning Medium	Cleaning Equipment
Plant floors	Most types of self-foaming or foam boosters added to most moderate to heavy-duty cleaners	Foam (high-pressure, low-volume should be used with heavy fat or protein deposits)	Portable or centralized foam cleaning equipment with foam guns for air injection into the cleaning solution
Plant walls and ceilings	Same as above	Foam	Same as above
Processing equipment and conveyors*	Moderate to heavy-duty alkalies that may be chlorinated or nonalkaline	High-pressure, low-volume spray	Portable or centralized high-pressure, low-volume equipment; sprays should be rotary hydraulic
Closed equipment	Low-foam, moderate to heavy-duty chlorinated alkalies with periodic use of acid cleaners as follow-up brighteners and neutralizers	CIP	Pumps, fan or ball sprays, and CIP tanks

*Packaging equipment can be effectively cleaned with gel cleaning equipment
Source: Stenson and Forwalter (1978).

Sanitizers

Workers should use sanitizers after cleaning to destroy microorganisms. Chapter 6 gives more information on sanitizing. The dairy industry uses steam, hot water, and chemical sanitizers most often.

Steam sanitizing. Steam needs to be in contact with a surface for a particular amount of time in order to destroy the microorganisms. For example, steam can sanitize assembled equipment when the surfaces are exposed for 15 minutes and the temperature of the water (condensed steam) leaving the equipment is 80°C (176°F). The usefulness of this method is limited, because it is difficult to maintain the right temperature and because the energy costs are high. Steam can also be more dangerous than other sanitizing methods, so it is not usually recommended.

Hot-water sanitizing. This system pumps hot water through the assembled equipment to heat the surfaces it touches to a given temperature for a specified time. Five minutes of contact with hot water can sanitize equipment when the water is at 80°C (176°F) as it comes out of the equipment. This technique is expensive because the energy costs are high.

Chemical sanitizing. This system pumps a sanitizer such as a halogen (usually chlorine or iodine compounds) through the assembled equipment for at least 1 minute. This technique requires that the sanitizer come in contact with all of the product surfaces. Because the sanitizer needs to touch every surface, it is important to use an appropriate method to apply the sanitizer in dairy operations.

In large-volume mechanized operations, sanitarians can apply the sanitizer through sanitary pipelines by *circulation* or pumping of a sanitizing solution through the system. Workers prepare the appropriate amount of sanitizing solution to be pumped through the system. The system should have slight back-pressure to make sure the sanitizer reaches the upper inner surface of the pipeline.

Small operations that cannot justify a mechanized system can sanitize equipment, utensils, and parts by submersion in the sanitizer solution. Workers usually need to submerge items for about 2 minutes and then drain and air-dry them on a clean surface.

Sanitarians can sanitize closed containers such as tanks and vats by *fogging.* They should mix the sanitizing solution at double the strength normally used and should expose the equipment for at least 5 minutes.

If sanitarians *spray* the sanitizer onto equipment, it must touch and completely wet all surfaces. The solution strength should be twice that normally used.

If the plant does not have mechanized sanitizing equipment, sanitarians can sanitize large open containers, such as cheese vats, by *brush application.* The brush should touch all areas. Therefore, this method has high labor costs.

Workers should not rinse sanitized surfaces with water, because this can recontaminate them with aerobic microorganisms. Workers should also protect sanitized surfaces from recontamination from other sources.

Cleaning Steps

Dairy operations require eight cleaning procedures:

1. *Cover electrical equipment* with polyethylene film or its equivalent.
2. *Remove large debris* during the production shift or before prerinsing.
3. *Disassemble equipment* as needed.
4. *Prerinse* to remove up to 90% of the soluble materials, loosen tightly bound soils, and help the cleaning compound penetrate in the next cleaning step.
5. *Apply cleaning compound(s)* using the proper cleaning equipment.
6. *Postrinse* to wash away soil and cleaning compounds and prevent soil from redepositing on the cleaned surface.
7. *Inspect* to make sure that the area and equipment are clean, and correct any problems.
8. *Sanitize* to destroy microorganisms. Then the area and equipment are less likely to contaminate processed products.

Less soil builds up on equipment when the workers heat products for the minimum amount of time at the minimum temperature possible, cool product heating surfaces before and after emptying processing vats, keep soil films moist by rinsing away foam and other debris with water at 40 to 45°C (104–113°F), and leave water in the processing vats until cleaning.

Other Cleaning Methods

When mechanized cleaning is not practical, workers should hand-clean equipment and surfaces, following these guidelines:

- Prerinse with water at 37 to 38°C (99–100°F).
- Use a cleaning compound with pH less than 10 to minimize skin irritation. The temperature of the cleaning solution should be 45°C (113°F). Solution-fed brushes work well for hand cleaning. Parts that are difficult to clean should be cleaned with cleaning-out-of-place (COP) equipment to remove lubricants and other deposits.
- Postrinse with water at 37 to 38°C (99–100°F) and air dry.
- Use a chlorine sanitizer as a spray or dip.

Table 11.2 describes hand cleaning of various types of dairy plant equipment.

CLEANING EQUIPMENT

Cleaning of dairy facilities involves removing soil from all surfaces that come in contact with food products and using a sanitizer after each processing period. Surfaces that do not touch food products also need to be cleaned. Dairy plants use different cleaning systems, depending on their size. A cleaning-in-place (CIP) system is used

TABLE 11.2. Special Considerations for Hand Cleaning Dairy Plant Equipment

Equipment	Recommended Cleaning Procedures
Weigh tanks (can receiving and/or in-plant can transfer)	Rinse immediately after milk has been removed. Disconnect and disassemble all valves and other fittings; wash weigh tank, rinse tank, and fittings; sanitize prior to next use.
Tank trucks, storage tanks, processing tanks	Remove outlet valve, drain, rinse several times with small volumes of tempered (38°C, 100°F) water, remove other fittings and agitator; brush or pressure-clean vats, tanks, and fittings; rinse and reassemble after sanitizing fittings just before reuse. Thoroughly clean manhole covers, valve outlets, sight glass recesses, and any air lines. High-pressure sprays are preferable to keep the cleanup personnel out of the tanks or vats and thus minimize damage to surface and contamination of cleaned surfaces.
Batch pasteurizers and heated product surfaces	Lower temperature below 49°C (120°F) immediately after emptying product; immediately rinse, with brushing to loosen burned-on products; if the vat cannot be rinsed, fill with warm (32–38°C, 85°F–100°F) water until cleaning. Clean the same as for other processing vats.
Coil vats	Although not in general use, they are difficult to clean because of inaccessibility of some surfaces of the coil. After prerinse, fill with hot water until bottom of coil is covered; add cleaning compounds and rotate coil while all exposed coil surfaces are brushed.
Homogenizers	Prerinse while the unit is assembled; dismantle and clean each piece; place cleaned parts on a parts cart to dry. Sanitize and reassemble prior to use.
Sanitary pumps	After use, remove head of pump, and flush thoroughly with tempered (38°C, 100°F) water; remove impellers and place them in the bucket containing a cleaning solution of 49–50°C (120°F–122°F). Wash intake and discharge parts and chamber. Brush impellers, and place them in a basket or on a parts table to dry.
Centrifugal machines	Non-CIP types must be cleaned by hand. Rinse with 38°C (100°F) water until discharge is clear. Dismantle, remove bowl and discs, and rinse each part before placing in the wash vat. A separate wash vat is desirable for separator and clarifier parts to avoid damaging the discs and other close-tolerance parts. Each disc should be washed separately, rinsed, and drained thoroughly. If a separator is used intermittently during the day, it should be rinsed after each use with at least 100 L (26 gal) of tempered water. Use of a mild alkaline wetting agent can improve rinsing efficiency.

by most large-volume plants. CIP is the best way to clean pipelines, milking machines, bulk storage tanks, and most processing equipment. Most dairy-processing plant equipment is used for less than 24 hours at a time, so the equipment and the area are cleaned daily. Longer continuous use of piping and storage systems allows workers to clean them less often, perhaps once every 3 days.

CIP and Recirculating Equipment

The effectiveness of CIP depends on time, temperature, concentration, and force. Rinse and wash times should be as short as possible to conserve water and cleaning compounds, but should be long enough to remove the soil. The time required for proper cleaning is affected by the temperature, concentration, and force of the cleaning solution. An energy-efficient CIP system can reduce cleaning costs by more than 35% and can use about 40% less energy.

A salmonellosis outbreak was caused by a batch of pasteurized milk in the 1980s because of a CIP cross-connection between raw and pasteurized products. Since this incident, many dairies have installed completely separate CIP systems for the receiving area of the plant.

The temperature of the cleaning solution used in CIP equipment should be as low as possible, but it should clean properly using as little cleaning compound as possible. The rinse temperature should be low enough to avoid hard-water deposits.

Force and physical action affect how well the cleaning compound reaches all areas of the equipment. Good equipment design and use of the right high-pressure pumps make sure that the force (or physical action) is adequate for efficient cleaning.

Although many types of spray devices are utilized in the dairy-processing industry, permanently installed fixed-spray units last longer than portable equipment and rotating or oscillating units. Fixed-spray units have no moving parts, stainless-steel construction, and fewer problems with performance if the pressure changes slightly.

In CIP, the cleaning compound must hit the surface with plenty of force, and workers must continuously replenish the supply of cleaning compound. Various CIP equipment systems are available (see Chap. 7). Some CIP systems now use water from the final rinse to make up the cleaning solution for the following cleaning cycle to cut down on the amount of wastewater.

COP Equipment

The following steps are recommended when using COP equipment in dairy plants:

1. Prerinse with water at 37 to 38°C (98–100°F) to remove heavy soil.
2. Wash by circulating a chlorinated alkali cleaning solution for approximately 10 to 12 minutes at 30 to 65°C (86–149°F) to loosen and get rid of the rest of the soil.
3. Postrinse with water at 37 to 38°C (98–100°F) to remove any soil or cleaning compound residues.

Cleaning Storage Equipment

Storage tanks need good spray devices to allow effective spray cleaning. Fixed-base, permanently installed sprays are more common in the industry than rotating and oscillating sprays. These sprays require less maintenance, are made of stainless steel, have no moving parts, and are durable. Their performance is not affected by slight changes in supply pressure, and the spray hits all of the surfaces continuously. Sprays can clean cylindrical and rectangular tanks properly. The spray patterns are designed to reach the upper one-third of the storage container. Because the equipment contains heating or cooling coils with complex agitators, the equipment needs a special spray pattern and higher pressure and volume to cover all of the surfaces.

Large tanks with spray devices prerinse and postrinse using a spray technique with three or more water bursts of 15 to 30 seconds each, with complete draining of the tank between bursts. This removes sedimented soil and foam better and uses less water than continuous rinsing.

The soil deposited in storage tanks and processing vessels varies more than that in piping circuits. Therefore, cleaning techniques for this equipment also vary more. The system for cleaning lightly soiled surfaces, such as those in storage tanks for milk or milk by-products, involves a three-burst prerinse of water at a controlled temperature, recirculation of a chlorinated alkaline detergent for 5 to 7 minutes at 55°C (131°F), a two-burst postrinse at tap water temperature, and recirculation of an acid final rinse for 1 to 2 minutes at tap water temperature. Recirculation times and temperatures may be longer for thicker products with more fat and total solids.

Soil on cold surfaces is different from burned-on deposits, which contain more protein and minerals. Burned-on soil requires a more concentrated cleaning compound solution, temperatures up to 82°C, and application for up to 60 minutes. Heavy burned-on deposits may also need a hot alkaline detergent and a hot acid detergent solution with mechanical action for removal.

Table 11.3 lists usual concentrations of cleaning compounds and sanitizers for various cleaning jobs.

Cleaning programs are tailored to the food product. Here are two examples of products that need special approaches:

Milk, skim-milk, and low-fat-milk-product–processing equipment: Because the mineral content of these products is high, the best way to clean the equipment is recirculation of an acid detergent for 20 to 30 minutes, followed by recirculation of a strong alkaline cleaner for approximately 45 minutes. The system may use a cold-water rinse between the acid and alkaline cleaners.

Cream and ice-cream–processing equipment: These products contain more fat and less minerals than milk products. The best way to clean this equipment is by recirculating an alkaline cleaner (0.5–1.5% strength) for approximately 30 minutes and adding an acid to produce a pH of 2.0 to 2.5. It is best to keep the temperature of the cleaning solution about 5°C (9°F) higher than the highest processing temperature during production.

TABLE 11.3.   Typical Concentrations for Various Cleaning Applications

Cleaning Application	Concentration, ppm	
	Chlorinated Cleaning Compounds*	Acid/Acid Anionic Chlorine Sanitizers
Milk storage and transportation tanks	1,500–2,000	100
Cream, condensed milk, and ice-cream storage tanks	2,500–3,000	100–130
Processing vessels for moderate heat treatment	4,000–5,000	100–200
Heavy "burn-on"	0.75–1.0% (causticity)	Acid wash at pH 2.0–2.5

*An acid rinse after cleaning should also be considered.

SUMMARY

- The layout and construction of dairy-processing plants affect sanitation and wholesomeness of the food products.
- The pathogen that causes most problems in dairy products is *L. monocytogenes*. This pathogen grows at refrigerator temperatures, so good sanitation is especially important.
- The plant's air and water supplies must be clean. All the surfaces in the plant, whether they come in contact with the products or not, must be easy to clean and sanitize.
- Dirt in dairy plants includes minerals, proteins, lipids, and carbohydrates from the milk and dairy products, as well as dust, lubricants, microorganisms, and residues of cleaning compounds and sanitizers.
- Good sanitation practices keep the amount of soil to a minimum.
- Sanitarians should choose the best combination of cleaning and sanitizing compounds and equipment for each cleaning job. Hot and cold surfaces have different types of soil and need different cleaning methods.
- If the water supply is hard, sanitarians should decide whether to soften it or overcome the problem with additions to cleaning compounds.
- CIP systems work well in many dairy-processing plants. Newer systems recycle rinse water and produce less wastewater.

BIBLIOGRAPHY

Campbell, J. R., and Marshall, R. T. 1975. *The Science of Providing Milk for Man.* McGraw-Hill Book Company, New York.

Hall, H. S. 1976. *Standardized Pilot Milk Plants.* Food and Agriculture Organization of the United Nations, Rome.

Heldman, D. R., and Seiberling, D. A. 1976. Environmental sanitation. In *Dairy Technology and Engineering*, p. 272. AVI Publishing Co., Westport, Conn.

Jowitt, R. 1980. *Hygienic Design and Operation of Food Plant.* AVI Publishing Co., Westport, Conn.

Marriott, N. G. 1994. *Principles of Food Sanitation*, 3d ed. Chapman & Hall, New York.

Marriott, N. G. 1990. *Meat Sanitation Guide II.* American Association of Meat Processors and Virginia Polytechnic Institute and State University, Blacksburg.

Seiberling, D. A. 1987. Process/CIP engineering for product safety. In *Food Protection Technology*, p. 181. Lewis Publishers, Chelsea, Mich.

STUDY QUESTIONS

1. Which microorganism causes the most problems in dairy foods? Give three reasons why.
2. What should processors think about when they choose where to build a dairy-processing plant?
3. What type of soil needs to be cleaned in dairy-processing plants?
4. What should sanitarians do if the water source is hard?
5. Why are expensive cleaning compounds sometimes the most economical?
6. What can processors do to reduce the amount of soil on heated surfaces?

TO FIND OUT MORE ABOUT DAIRY PROCESSING AND PRODUCT SANITATION

1. Contact the National Dairy Council, (708) 803-2000, ext. 220, and ask for a catalog of their education materials.
2. Contact the Food Marketing Institute, (202) 429-8298 or (202) 429-8266, and ask for a copy of their brochure, *A Consumer Guide to Food Quality & Safe Food Handling: Dairy Products & Eggs.*
3. Contact a dairy-processing plant in your area (one that processes milk or makes cheese or yogurt), and ask them how they handle sanitation.
4. Contact your local inspector. Ask for information on the inspection program. Ask if you can "shadow" an inspector (with approval of the plant manager) during an inspection.

WHY ALL THE FUSS ABOUT *Listeria monocytogenes?*

- *Listeria monocytogenes* can survive in aerobic and anaerobic environments, so it can live in several different types of food.
- *L. monocytogenes* can be found in animal intestines, green plants, water, and sewage.
- *L. monocytogenes* can grow and multiple at refrigeration temperatures.
- Pregnant women are encouraged to eat and drink plenty of dairy products, which are high in calcium and other nutrients. Listeriosis can cause spontaneous abortion or stillbirth of the fetus in pregnant women.
- Listeriosis is most common in newborn babies, the elderly, people with a weakened immune system (e.g., people with AIDS), or people with other diseases.
- Of the people who get listeriosis, 25% die.
- If milk is heavily contaminated with *L. monocytogenes,* it can survive minimum HTST (high-temperature, short-time) pasteurization, although it is normally destroyed by pasteurization and processing.
- Dairy products can be contaminated by air currents, dirty filters in ventilation systems, water (condensate from moisture in the air), and plant staff.
- Dairy-processing plants contain many sources of *Listeria* contamination. Excellent sanitation is essential to control this pathogen.
- Scientists do not know what level of contamination is needed to cause infection. Therefore, the U.S. Food and Drug Administration does not tolerate *any* detectable *L. monocytogenes* microbes.

Source: Donnelly, C. W. 1990. Concerns of microbial pathogens in association with dairy foods. *J. Dairy Sci.* 73:1656.

CHAPTER 12

Meat and Poultry Sanitation

ABOUT THIS CHAPTER

In this chapter you will learn:

1. Why it is so important to have good sanitation in meat- and poultry-processing facilities
2. Which pathogens cause most problems in meat and poultry food products
3. How to design and organize meat and poultry plants to control pathogens
4. How to clean and sanitize the various parts of meat- and poultry-processing plants.
5. How HACCP is important in meat- and poultry-processing plants

INTRODUCTION

Meat and poultry are perishable foods (i.e., they spoil quickly), and the meat discolors easily. Poor sanitation allows microbes to cause even more damage to the flavor and color of meat. A good sanitation program reduces color changes and spoilage and increases the shelf life of meat and poultry.

In the meat and poultry industry, sanitation begins with the live animal or bird and continues until the food product is served. Management should plan, enforce, and supervise the sanitation program thoroughly. The person in charge of the program should report to top management and make sure that the facility and food products are sanitary. Trained staff should clean and inspect the plant and equipment and be responsible for keeping it clean.

COMMON PATHOGENS

Meat and poultry products cause about 23% of foodborne illness outbreaks. They are associated with 5 to 10% of deaths caused by foodborne disease. Pathogens associated with meat and poultry are *Listeria monocytogenes*, *Escherichia coli* O157:H7, *Salmonella* spp., and *Campylobacter* spp.

Listeria monocytogenes is often found on retail cuts of fresh poultry and red meats. *L. monocytogenes* can also grow in some packaged cooked meat products, such as hotdogs. It is important to make sure that fresh meats used to make processed products do not recontaminate ready-to-eat products after they have been processed. Table 12.1 shows how often *L. monocytogenes* was found in various places in 41 meat plants.

TABLE 12.1. Percentage of Meat Plants That Tested Positive for *L. Monocytogenes* in Various Areas

Location	Plants, %
Floors	37
Drains	37
Cleaning aids	24
Wash areas	24
Sausage peelers	22
Food contact surfaces	20
Condensation	7
Walls and ceilings	5
Compressed air	4

Source: Wilson, G. D. 1987. Guidelines for production of ready-to-eat meat products. *Proc. Meat Ind. Res. Conf.,* p. 62. American Meat Institute, Washington, D.C.

L. monocytogenes is often found around wet areas and cleaning equipment, such as floors, drains, wash areas, ceiling condensation, mops and sponges, brine chillers, and at peeler stations. Refrigeration at 4 to 5°C (39–41°F) (a common storage temperature) does not stop this pathogen from growing. Possible ways to control *L. monocytogenes* include antimicrobial agents, colder storage (<2°C/<35°F), making products with lower water activity or pH, or pasteurizing products after processing.

One of the best ways to control pathogens like *L. monocytogenes* is by preventing cross-contamination. Staff who work in areas that contain both raw and finished product, such as smokehouses and water- and steam-cooking areas, should change their outer clothing and sanitize their hands or change gloves when moving from raw- to finished-product areas. Staff should sanitize utensils and thermometers that are used for raw and finished products each time they use them. Staff must also scrub floors frequently. If liquid condenses on the ceiling, staff should remove it using a vacuum cleaner or sanitized sponge mop.

SANITATION MANAGEMENT

Meat and poultry are ideal food for microorganisms that cause discoloration, spoilage, and foodborne illness. Good sanitation keeps contamination to a minimum and increases the stability of the product.

Some of the most important reasons for having high standards of cleanliness in meat and poultry facilities are:

- When conditions are unsanitary, microorganisms that attack meat and poultry are present. These microorganisms change the color and flavor of products.
- Self-service packages of fresh meat and poultry in stores make sanitation very important to extend the shelf life as long as possible.
- Good sanitary conditions reduce the amount of waste, because less-discolored or spoiled meat and poultry have to be thrown out.
- Excellent sanitary conditions can improve the image and reputation of a firm. A sanitary product looks and tastes better and is more healthful than tainted food.
- Regulatory agencies and consumers insist on good sanitation.
- Employees deserve clean, safe working conditions. Sanitary and tidy surroundings improve morale and productivity.
- More centralized processing and packaging requires more emphasis on sanitation. More processing and handling of food require a stricter sanitation program.
- Sanitation is good business.

Microorganisms are a major cause of color changes in meat and poultry. Microbes take up oxygen from the surface of the meat, so there is less oxygen to keep the muscle color bright. Different microorganisms have different effects on color pigments. But good cleanliness can slow the growth of microbes. Employees who handle meat should try to minimize the number of bacteria that first come in contact with the meat.

Meat and Poultry Contamination

During slaughter, processing, distribution, and foodservice, meat and poultry items may be handled as many as 18 to 20 times. Almost anything that touches meat and poultry can contaminate it, so the risk of contamination increases with the number of times the product is handled.

A healthy, live animal has defenses (immune system or resistance) that stop bacteria from getting into its muscles. After slaughter, the natural defenses stop working, and there is a race between human beings and the microbes to see which will eat the meat! If processors handle the meat carelessly, the microbes win. Sanitation employees have to find ways to stop the microorganisms from growing. (Chapter 3 discusses contamination sources during slaughter and processing.)

Layout and Plant Design

1. The plant layout should keep pests out and stop *L. monocytogenes* from moving between raw- and cooked-product areas. Designers need to consider employee traffic patterns, where support and supervisory staff work, and where and how food is handled.
2. Designers should use air and refrigeration equipment that is easy to clean and sanitize. Air should not flow from raw-food areas to cooked-food areas.
3. All equipment and other surfaces should have smooth, nonporous surfaces that are easy to clean and sanitize.
4. Floor surfaces should be easy to clean and should not collect water.
5. The design should prevent condensation from collecting.

Equipment Design

1. The facility should use only regulatory approved materials.
2. Staff should maintain and store all equipment properly.

Process Control

1. If the process does not contain a step to kill *L. monocytogenes,* the operation should be designed to minimize contamination.
2. The step to kill *L. monocytogenes* (if there is one) should be a critical control point in the Hazard Analysis Critical Control Point (HACCP) program.

Operation Practices

1. Management should educate employees about their responsibility for using good manufacturing practices (GMPs) and HACCP.
2. Employees should have the right sanitation equipment, such as footbaths, hand dips, hairnets, and gloves.
3. Contamination sources should be removed, especially in areas where food is processed and ready to eat.
4. Management should understand and support GMPs and HACCP.

Sanitation Practices

1. The operation should have an adequate number of employees, amount of time, and number of supervisors for cleaning and sanitizing.
2. The facility should have written cleaning and sanitizing procedures posted for each area.

3. The environment (equipment, walls, water, air, etc.) should be sampled to make sure that cleaning and sanitizing procedures are working.

Checking Control of Pathogens

1. Weekly samples should be taken from plant areas, equipment, and the air supply for microbial counts. It is especially important to take samples from points between the step used to kill *L. monocytogenes* and packaging.
2. Composite (combined) samples are cheaper to analyze than individual samples. If a composite sample shows pathogens, sanitation staff need to go back and analyze the individual samples to find out which one is contaminated.

The following steps can help control pathogens:

1. Mechanically or manually scrub floors and drains every day. Rinse drains with disinfectant every day.
2. Scrub walls each week.
3. Clean the outside of all equipment, light fixtures, sills and ledges, piping, vents, and other processing and packaging areas that are not part of the daily cleaning program each week.
4. Clean cooling and heating units and ducts each week.
5. Caulk all cracks in walls, ceilings, and window and door sills.
6. Scrub and clean raw-material areas at least as often as the processing and packaging areas.
7. Keep hallways and passageways used by both raw and finished product clean and dry.
8. Minimize traffic in and out of processing and packaging areas. Establish plant traffic patterns to reduce cross-contamination from feet, containers, pallet jacks, pallets, and fork trucks.
9. Change outer clothing and sanitize hands or change gloves when moving from an area containing raw ingredients to an area containing finished food products.
10. Change into clean work clothes each day at the plant. Color-code clothing to show which clothes belong in each area.
11. Make sure visitors change into clean clothes at the plant.
12. Monitor the plant environment (surfaces and air) to make sure that *Listeria* are under control.
13. Close-in processing and packaging rooms so that they only receive filtered air.
14. Clean and sanitize all equipment and containers before bringing them into processing and packaging areas.

Temperature Control

Meat and poultry spoil at high temperatures. Chemical and biochemical reactions are faster and the lag phase of the growth pattern of microorganisms is shorter at higher temperatures. Microbial and nonmicrobial spoilage both increase at higher temperatures until 45°C (113°F). Above 60°C (140°F), microbes do not usually spoil food. (See Chap. 2 for more about microbial growth.)

Microorganisms grow fastest between 2°C and 60°C (36–140°F). This temperature range is the critical zone, or the danger zone. Meat and poultry must be stored either below or above this temperature zone and should be taken through this range as quickly as possible when a temperature change is necessary (e.g., cooking or chilling). Storage at temperatures below the critical zone (refrigerator or freezer) does not kill bacteria, but they can only grow and multiply very slowly. Below the critical zone, bacteria are less active and some may die through stress (see Fig. 12.1).

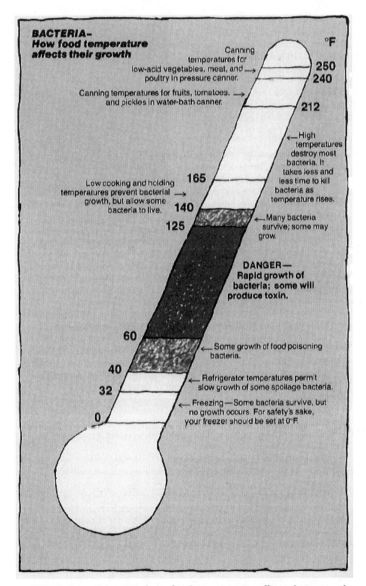

FIGURE 12.1. Bacteria—how food temperature affects their growth. (*Source: The Safe Food Book: Your Kitchen Guide, p.8, Home & Garden Bulletin No. 241, United States Department of Agriculture, Food Safety and Inspection Service, July 1984.*)

Processing and storage at colder temperatures reduces spoilage and also reduces growth of microorganisms on equipment, supplies, or other materials in the storage and processing areas. If conditions are unsanitary and the temperature is not controlled, some types of *Pseudomonas* can double in number every 20 minutes. This means that in 12 hours, 1 bacterium can become 280 trillion bacteria. At 0°C (32°F), meat and poultry have twice as long a shelf life as meat and poultry at 4.4°C (40°F) and at least 4 times as long a shelf life as meat and poultry at 10°C (50°F).

Air curtains can prevent cold air from leaving an area, especially if truck doors have to be left open, and keep out insects and dust. The air stream should go across the entire width of staff entrances.

It is important to clean the roofs above food-manufacturing areas. Equipment and exhaust stacks may be vented through the roof. If possible, designers should enclose processing equipment that hangs from the ceiling with a floor to separate it from the processing area. Particles can deposit on the outside of the roof, especially if it is flat, and can attract birds, rodents, or insects. Pools of water also encourage pests. The slope of the roof should be at least 1% for drainage.

CLEANING PRACTICES FOR MEAT AND POULTRY PLANTS

An efficient cleaning system can reduce labor costs up to 50%. Proper construction and cleaning equipment are critical for a good cleaning operation. Floors, walls, and ceilings should be impervious (nonporous) and easily cleaned. Floors should have a slope of at least 10.5 mm/m (0.4 in/yd).

Hot-Water Wash

Soil from meat and poultry is mostly fat and protein, so a hot-water wash does not work well. Hot water can loosen and melt fat, but it also tends to cook fats and proteins onto the surface. The main advantage of a hot-water-wash system is that it does not need much equipment. But it uses a lot of labor and causes water condensation on equipment, walls, and ceilings. It is difficult to remove heavy soil using a hot-water wash.

High-Pressure, Low-Volume Cleaning

High-pressure, low-volume spray cleaning works well in the meat and poultry industry because it removes soil well. With this equipment, the operator can clean difficult-to-reach areas with less labor, and the cleaning compound works better at a cooler temperature.

Portable high-pressure, low-volume units can be easily moved around the plant to clean parts of equipment and building surfaces. It works especially well for conveyors and processing equipment where soaking is impractical and hand brushing is difficult and labor intensive.

Chapter 7 includes illustrations of the metering device and controls of a centralized high-pressure cleaning unit. Figure 12.2 shows a dispensing nozzle for this equipment.

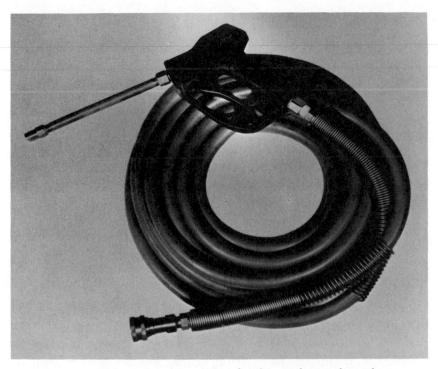

FIGURE 12.2. High-pressure hose with a female, stainless-steel, quick-connect, heavy-duty, dead-man shutoff-type spray gun, extension wand, and 4- to 12-liter-per-minute (1–3 gallon-per-minute) capacity spray nozzle. *(Courtesy of Klenzade Division, a service of Ecolab.)*

A centralized high-pressure cleaning system has many uses and benefits:

- The equipment can dispense several types of cleaning compound—acid, alkaline, or neutral cleaners and sanitizers.
- The system can use mechanized spray heads, mounted on belt conveyors with automatic washing, rinsing, and cutoff.
- Quick connection outlets are available in all areas requiring cleaning.

Centralized equipment is custom built and is far more expensive than portable units, depending on the size of the facility and the flexibility of the equipment. The initial cost may range from $15,000 to over $150,000.

Foam Cleaning

Foam is useful for cleaning large surface areas of meat and poultry plants, the outside of transportation equipment, ceilings, walls, piping, belts, and storage containers. Portable foam equipment is similar in size and cost to portable high-pressure units. Centralized foam cleaning has the same benefits as other centralized systems.

Gel Cleaning

This equipment is similar to high-pressure units, except that the cleaning compound is applied as a gel rather than as a high-pressure spray. Gel is good for cleaning packaging equipment because it clings to soiled surfaces. The equipment costs about the same as portable high-pressure units.

Combination Centralized High-Pressure, Low-Volume, and Foam Cleaning

This system is the same as a centralized high-pressure system, except that the system can also apply foam. This method is flexible because foam can clean large surfaces and high pressure can clean belts, conveyors, and hard-to-reach areas. This system costs from $15,000 to over $150,000, depending on the size of the system.

Cleaning in Place (CIP)

Nozzles inside the equipment apply a recirculating cleaning solution to automatically clean, rinse, and sanitize the equipment. The benefits of CIP systems are discussed in Chapter 7. CIP cannot be used much in the meat and poultry industry. The equipment is expensive and does not work well in heavily soiled areas. CIP cleaning is sometimes used to clean vacuum thawing chambers, pumping and brine circulation lines, silos, and fat-rendering systems. Figure 12.3 illustrates a CIP system for

FIGURE 12.3. Shackle washer for cleaning shackles, rollers, and the chain in poultry-processing plants. *(Courtesy of Chemidyne Corp.)*

washing shackles, rollers, and the chain in poultry plants. The motor and drive components are mounted on a base plate. The shackles are cleaned as they pass between two rotating brushes. The brushes can be lifted above the rail when not in use.

CLEANING COMPOUNDS FOR MEAT AND POULTRY PLANTS

Acid Cleaners

Information about strongly and mildly acid cleaners is provided in Chapter 5.

Strongly Alkaline Cleaners

Examples of strongly alkaline compounds are sodium hydroxide (caustic soda) and silicates with high ratios of nitrous oxide to silicon dioxide. Silicates make sodium hydroxide less corrosive and better at penetrating and rising. These cleaners remove heavy soils, such as those found in smokehouses.

Heavy-Duty Alkaline Cleaners

These cleaners are frequently used with CIP, high-pressure, and other mechanized systems in meat and poultry plants.

Mild Alkaline Cleaners

Mild cleaners are often liquid solutions and are used for hand cleaning lightly soiled areas in meat and poultry plants.

Neutral Cleaners

Information about these and other cleaning compounds is given in Chapter 5.

SANITIZERS FOR MEAT AND POULTRY PLANTS

Sanitizers work only on clean surfaces, i.e., after all dirt is removed. Soil that can prevent the sanitizer from working properly includes fats, meat juices, blood, grease, oil, and mineral buildup. Microbes can grow under and inside soil and can hold food and water that allow microbes to grow. Chemical sanitizers cannot destroy microorganisms in or under soil deposits.

Steam

Steam is not a good sanitizer. Many operators think that water vapor is steam and do not expose the equipment to enough steam to sanitize it. Workers should not use steam to sanitize refrigerated areas because it causes condensation and wastes energy. Steam also does not sanitize conveyors properly.

Chemical Sanitizers

Chlorine is the most important compound for disinfecting, sterilizing, and sanitizing equipment, utensils, and water. Meat and poultry operations most often use these chlorine compounds:

Sodium and calcium hypochlorite: These cost more than plain chlorine, but are easier to apply.

Liquid chlorine: This chlorinates processing and cooling waters and prevents bacterial slimes.

Chlorine dioxide: This works well to destroy bacteria when organic matter (such as food waste) is present because it does not react with protein.

Active iodine solutions can also sanitize. Iodophors are very stable, have a much longer shelf life than hypochlorites, and work at low concentrations. These sanitizing compounds are easy to measure and dispense, and penetrate dirt well. They prevent films and spotting because of their acidity. The temperature of the sanitizing solution should be below 48°C (118°F).

Quaternary ammonium compounds work well on floors, walls, equipment, and furnishings in meat and poultry plants. The "quats" work well on porous surfaces because they penetrate well. Quats form a film on surfaces which inhibits bacterial growth. Sanitizers and compounds that contain an acid and a quat sanitizer work best for controlling *Listeria monocytogenes* and mold. Sanitarians may use quats temporarily when they find mold buildup.

Acid sanitizers combine rinsing and sanitizing steps. Acid neutralizes the excess alkali from the cleaning residues, prevents alkaline deposits from forming, and sanitizes. Chapter 6 has more information about sanitizers.

PERSONAL HYGIENE AND WORK HABITS OF EMPLOYEES

General Instructions

All personnel should have good personal hygiene (see Chap. 4). They should wear clean clothes and stay away from work if they are ill. Only supervisors, managers, and superintendents should have access to cleaning and sanitizing compounds and should be responsible for giving them to cleaning staff. Wrong use of these compounds is expensive, does not result in good cleaning, and could damage employees or equipment. Supervisors should lock the water temperature at 55°C (131°F).

Instructions for portable or centralized high-pressure or foam cleaning systems should be followed. The vendor's recommendations for using cleaning compounds should be followed. (See Chap. 4 for information on safety precautions for cleaning compounds.)

The sanitation supervisor should inspect the entire plant each night after cleaning. All soiled areas should be cleaned again before the regulatory agency inspects the plant in the morning.

Chlorine test papers should be used to check the concentration of sanitizing solutions if automatic mixing equipment is not available. The test papers come in vials of 100 strips each and are available through most cleaning-compound suppliers.

Recommended Sanitary Work Habits

Workers should:

1. Store personal equipment (lunch, clothing, etc.) in a sanitary place, and always keep storage lockers clean.
2. Wash and sanitize utensils frequently throughout the production shift, and store them in a sanitary container that will not be in contact with floors, clothing, lockers, or pockets.
3. Not allow food to touch surfaces that have not been sanitized. If any food touches the floor or other unclean surface, it should be rinsed thoroughly with drinkable water.
4. Use disposable towels for hands or utensils.
5. Wear clean clothing in production areas.
6. Cover their hair so that it does not contaminate food.
7. Remove their aprons, gloves, or other clothing that may touch the product before using toilets.
8. Always wash and sanitize their hands when leaving the toilet area.
9. Stay away from production areas when they have an infectious disease, infected wound, cold, sore throat, or skin disease.
10. Not use tobacco in any production area.

HACCP

In the future, the U.S. Department of Agriculture, which regulates the hygienic production of meat and poultry products, may require use of a HACCP program. (Chapter 10 gives more detail about HACCP.)

HACCP does not have to be expensive. The control points are chosen because they are important for keeping the food safe. Design, maintenance, and process control are important control points and are quite inexpensive. Microbial tests are expensive and do not control the process well.

Sanitation staff can develop a simple HACCP flowchart for the meat or poultry production line. The flowchart includes a long sequence of events with steps that are difficult or impossible to control. Sanitation employees can identify the factors that affect the hazards of each step and decide on critical control points.

Components of Meat and Poultry Production for HACCP Analysis

Livestock and poultry production. It is possible to raise animals in a specific pathogen-free (SPF) environment to reduce contamination. Farmers can also give animals cultures of good bacteria to compete with pathogens and keep their numbers down.

Farm environment. The farm (pastures, streams, manure, etc.) helps recycle excreted pathogens in a cycle of infection, excretion, and reinfection. Sanitation practices can improve hygiene in this area.

Transportation. Transporting live animals is stressful and may cause animals carrying pathogens to spread these microorganisms. It is challenging to keep transportation sanitary to reduce contamination in the processing plant.

Animal-holding pens. Stress in animal-holding pens can cause changes in the balance of microbes in the intestines so that the animals shed more *Salmonella* spp. in their feces. Showering the animals can reduce the amount of stress and contamination.

Hide, pelt, hair, or feather removal. Meat animals have protective coats that often contain harmful microorganisms. The industry needs better procedures and equipment to remove the outer layers without contaminating the meat.

Removal of internal organs. Intestines can spill and internal organs can burst, releasing microbes onto the meat. A series of water or sanitizer sprays can reduce contamination of poultry or red-meat carcasses. Spraying does not completely remove microorganisms and can spread them over the carcass.

Inspection. A meat inspector can contaminate dressed carcasses with his or her hands or a knife. The inspector should sanitize the knife and his or her hands.

Chilling. Air temperature, air movement, relative humidity, and filtering air all affect the speed of chilling. Keeping the carcass surface dry helps stop microorganisms from growing. Trimming the neck flap area of poultry carcasses after chilling can reduce the number of microbes.

Further processing. Processors should not allow chilled carcasses and cuts of meat to get warm. Processing equipment should have a hygienic design, and should be sanitized before use. Any ingredients added to processed meats should be safe and wholesome.

Packaging. Proper packaging protects the food product from contamination. Proper storage temperatures are also important.

Distribution. Distribution must be fast, clean, and cold. Vehicles should be inspected for cleanliness and temperature control.

Critical Control Points for Beef Production

The following critical control points (CCPs) have been suggested by the Meat and Poultry Group of the National Committee on Microbiological Criteria for Foods for reducing the risk of pathogens on fresh beef:

CCP1-Skinning: The hide is very contaminated, and there is no good way to remove soil from the live animal before slaughter. Therefore, it is important to skin the animal to cut down on cross-contamination from the hide to the carcass.

CCP2-Prewash: This step should remove most of the contamination from contact with the hide and minimize attachment of microbes to the carcass.

CCP3-Bacterial wash: Bactericidal compounds (e.g., acetic acid) reduce contamination with microbes, including pathogens from the animal's intestines.

CCP4-Evisceration: Animals hold large numbers of pathogens in their intestines. The best way to make sure these pathogens do not contaminate the carcass is to teach operators how to remove the intestines and internal organs without bursting them.

CCP5-Final wash: This bactericidal rinse reduces the number of microbes on the carcass and minimizes the number of pathogens that are carried into the rest of the processing and packaging process.

CCP6-Chill: Rapid chilling controls the growth of pathogens.

CCP7-Storage: Strict temperature control and daily cleaning and sanitizing of equipment prevent pathogens from growing. The population of pathogens at this point shows how well the previous CCPs have controlled contamination. Maintaining the product temperature below 7°C (45°F) is a CCP to prevent growth of pathogens during packaging and distribution.

SANITATION PROCEDURES

It is a good idea to have written cleaning instructions for each operation. Written procedures are especially useful when there is a change in supervisor and for training new employees. More mechanized processes have more detailed and complicated cleaning methods. Written details of cleaning operations should be posted in the plant.

Before deciding on a cleaning procedure, sanitarians must understand how all production and cleaning equipment operates. Here are some examples of cleaning procedures for various areas of a plant. These examples are only guidelines. Sanitarians need to adapt cleaning procedures to the particular area and process.

Livestock and Poultry Trucks

Frequency. After hauling each load.

Procedure
1. Immediately after removing livestock or poultry from trucks, scrape and remove all manure.
2. Clean the truck beds, wheels, and frame by washing down the racks, floors, and frames with water to completely remove all manure, mud, and other debris. Completely disinfect with a quaternary ammonium sanitizer spray, or clean and sanitize in one step by spray cleaning with an alkaline detergent sanitizer.

Livestock Pens

Frequency. As soon as possible after removing each lot.

Procedure

1. After taking the livestock from each pen, clean the manure from the floors and walls, and remove it from the plant.
2. Every 4 months, scrape all dried manure and loose whitewash off the gates and partitions. Sweep cobwebs from the ceilings, and whitewash the inside of the pens. Mix an acid-type sanitizer with the whitewash.
3. If any animals have an infectious disease, quarantine the diseased animals and destroy them separately from the healthy livestock. Remove the manure from the pen area (use a hose if necessary), and disinfect the pens by spraying with a quaternary ammonium sanitizer.

A general cleaning procedure for slaughter and processing areas includes:

1. Removing general debris
2. Prerinsing and wetting
3. Applying the cleaning compound
4. Rinsing
5. Inspecting
6. Sanitizing
7. Preventing recontamination

The first step saves time and water and cuts down the load that enters the sewage system. Removing debris also means less splashing of large particles during the second step.

Slaughter Area

Frequency. Every day. Debris should be removed several times during the production shift.

Procedure

1. Pick up all large pieces of debris and put into trash containers.
2. Cover all electrical connections with plastic sheeting.
3. Quickly prerinse all soiled areas with water at 50 to 55°C (122–131°F). Start with the ceiling, walls, and the top of the equipment, and keep debris moving down to the floor. Do not spray water directly onto motors, outlets, and electrical cables.
4. Apply an alkaline cleaner using a centralized or portable foam system and water at 50 to 55°C (122–131°F). The system should reach all framework, undersides, and other difficult-to-reach areas. Make sure the foam is in contact with the surfaces for 5 to 20 minutes before rinsing. Foam requires less labor, but high-pressure equipment is better for hard-to-reach areas and for removing *Listeria monocytogenes*.

5. Rinse ceilings, walls, and equipment no more than 20 minutes after applying the cleaning compound. Again, start at the ceiling and work downwards, and use water at 50 to 55°C (122–131°F).

6. Inspect all equipment and surfaces, and reclean if necessary.

7. Apply an organic sanitizer to all equipment using a centralized or portable sanitizing unit. If a chlorine sanitizer is used, the solution should be at least 50 parts per million (ppm) of chlorine.

8. Remove, clean, and replace drain covers.

9. Put a white edible oil on surfaces that may rust or corrode.

10. Clean specialized equipment according to the manufacturer's recommendations (if available).

11. Require maintenance workers to carry a sanitizer, and sanitize where they have worked so that equipment is not contaminated during maintenance and setup.

Poultry Mechanical Eviscerators

Frequency. Daily. Use a continuous or intermittent sanitizer spray to reduce contamination.

Procedure
1. Pick up all large pieces of debris and put into trash containers.
2. Cover electrical connections with plastic sheeting.
3. Quickly prerinse the equipment with water at 50 to 55°C (122–131°F).
4. Apply an alkaline cleaner using a centralized or portable foam system and water at 50 to 55°C (122–131°F). Expose the equipment to the cleaner for 10 to 20 minutes, and then rinse with water at 50 to 55°C (122–131°F).
5. Inspect all areas and reclean if necessary.
6. Apply a solution of chlorine at 200 ppm (or other organic sanitizer) using a centralized or portable sanitizing unit.
7. Use sanitizers to avoid contamination during maintenance as before.

Receiving and Shipping Area

Frequency. Daily.

Procedure
1. Cover all electrical connections, scales, and food product with plastic sheeting.
2. Quickly rinse the walls and floors with high-pressure water at 50 to 55°C (122–131°F). Begin rinsing the walls at the top, and rinse dirt down towards the floor. This prerinse removes heavy soil and wets the surfaces.
3. Apply an acid cleaner (55°C [131°F] or less) using a slurry or foam gun. Use high pressure (i.e., 25–70 kg/cm^2 [360–1,000 lb/in^2] and 7.5 to 12 L/min [2–3 gal/min] at the wand).

4. Apply the cleaning compound for 20 minutes, then rinse using high-pressure water at 50 to 55°C (122–131°F).
5. Remove, clean, and replace drain covers.
6. Put all hoses away.

Processed-Products Area

Frequency. Daily.

Procedure

1. Take all equipment apart and place the parts on a table or rack.
2. Pick up large pieces of meat and other debris and put into trash containers.
3. Cover all electrical connections with plastic sheeting.
4. Prerinse all surfaces with water at 50 to 55°C (122–131°F). Start at the top of processing equipment, and wash debris down to the floor. Do not spray motors, outlets, and electrical cables.
5. Apply an alkaline cleaner using a centralized or portable high-pressure, low-volume system using water at 50 to 55°C (122–131°F). Make sure the system reaches all framework, tables, the undersides of equipment, and other difficult-to-reach areas. Soak the equipment in the cleaning solution for 5 to 20 minutes. Foam can also be used, although it does not penetrate well.
6. Rinse using water at 50 to 55°C (122–131°F). Rinse one side of each piece of equipment at a time.
7. Inspect all equipment surfaces and reclean if necessary.
8. Apply an organic sanitizer to all clean equipment using a centralized or portable sanitizing unit.
9. Remove, clean, and replace drain covers.
10. Use a white edible oil on surfaces that may rust or corrode.
11. Be careful not to contaminate equipment during maintenance.

Fresh-Product Processing Areas

Frequency. Daily.

Procedure

1. Take all equipment apart and place the parts on a table or rack.
2. Remove large pieces of debris from equipment and the floor and put into trash containers.
3. Cover mixer, packaging equipment, motors, outlets, scales, controls, and fat analysis equipment with plastic sheeting.
4. Quickly prerinse all soiled surfaces with water at 50 to 55°C (122–131°F) to wash off large pieces of debris and to wet the surfaces. Hose debris towards a floor drain.
5. Apply an alkaline cleaner using centralized or portable high-pressure, low-volume cleaning equipment and water at 50 to 55°C (122–131°F). Or apply the

cleaning compound as a foam, gel, or slurry. Make sure the cleaning compound covers the whole area, including equipment, floors, walls, and doors.

6. Rinse the area and equipment within 20 to 25 minutes after applying the detergent.
7. Inspect the area and equipment and reclean if necessary.
8. Remove, clean, and replace drain covers.
9. Sanitize all clean equipment with an organic sanitizer using a centralized or portable sanitizing unit.
10. Use a white edible oil on surfaces that may rust or corrode.
11. Be careful not to contaminate equipment during maintenance.

Processed-Products Packaging Area

Frequency. Daily.

Procedure

1. Take all equipment apart and place the parts on a table or a rack.
2. Remove large pieces of debris from equipment and floors and put in trash containers.
3. Cover packaging equipment, motors, outlets, scales, controls, and other equipment with plastic sheeting.
4. Prerinse all soiled surfaces with water at 50 to 55°C (122–131°F) to remove heavy soil and to wet the surfaces. Hose debris towards a floor drain
5. Apply an alkaline cleaner using centralized or portable high-pressure, low-volume cleaning equipment and water at 50 to 55°C (122–131°F). Or use a foam, gel, or slurry to apply the cleaning compound. Make sure the cleaning compound covers the entire area, including equipment, floors, walls, and doors.
6. Rinse the area and equipment within 20 to 25 minutes after applying the cleaning compound.
7. Inspect the area and equipment and reclean if necessary.
8. Remove, clean, and replace drain covers.
9. Sanitize the clean equipment with an organic sanitizer using a centralized or portable sanitizing unit.
10. Use a white edible oil on surfaces that may rust or corrode.
11. Be careful not to contaminate equipment during maintenance.

Brine-Curing and Packaging Area

Frequency. Daily.

Procedure

1. Pick up all large pieces of debris and put in a trash container.
2. Cover all electrical connections, scales, and food products with plastic sheeting.

3. Prerinse the area and equipment with water at 50 to 55°C (122–131°F).
4. Put an acid cleaner in the shrink tunnel (if used), and circulate for about 30 minutes while prerinsing.
5. Rinse the shrink tunnel before applying the detergent.
6. Put all debris from prerinsing in a trash container.
7. Apply an alkaline cleaner using a foam or slurry cleaning system and water at 50 to 55°C (122–131°F).
8. Rinse with water at 50 to 55°C (122–131°F) no more than 20 minutes after applying the detergent.
9. Inspect the area and equipment and reclean if necessary.
10. Remove, clean, and replace the drain covers.
11. Sanitize all clean equipment using an organic sanitizer and a central or portable system.
12. Use white edible oil on parts that may rust or corrode.
13. Be careful not to contaminate equipment during maintenance.

Smokehouses

Frequency. After each smoke period.

Procedure
1. Pick up large pieces of debris and put in trash containers.
2. Apply an alkaline cleaning compound recommended for cleaning smokehouses using a centralized or portable foam system. Use a high-pressure unit where foam cannot penetrate. Figure 12.4 shows a unit for cleaning a smokehouse.
3. Rinse the area no more than 20 to 30 minutes after applying the cleaning compound. Start by rinsing the ceiling and walls, and work all debris down to the floor drain.
4. Inspect all areas and reclean if necessary.
5. Apply an iodophor or quaternary ammonium sanitizer using a centralized or portable sanitizing unit around the entry area to cut down on air contamination.

Smoke Generator

Frequency. Depends on amount of use.

Procedure
a. Filter
1. Soak the filter in an alkaline cleaning solution
2. If the filter has mineral deposits, cut the frame apart and clean each leaf. Reweld the frame after cleaning and avoid warping it.

FIGURE 12.4. A portable atomizer that covers all stainless-steel surfaces in a smoke-house and reduces cleaning time by 60 to 75%. *(Courtesy of Birko Chemical Corporation.)*

b. Wash Chamber

1. Take apart the duct connecting the smoke generator to the house.
2. Clean the soot and ash out of the chamber below the filter.
3. Clean the duct and chamber surface until the metal shows.

Wire Pallets and Metal Containers

Frequency. Before use.

Procedure

1. Use high-pressure water at 50 to 55°C (122–131°F) or cooler as a prerinse.
2. Apply an alkaline cleaner using a foam unit. If foam is not available, use a high-pressure, low-volume unit. Do not spray too many containers at one time. Be careful to rinse the containers before the cleaning compound dries.
3. Rinse with a high-pressure spray of water at 50 to 55°C (122–131°F).
4. Inspect all containers and reclean if needed.

Offices, Locker Rooms, and Rest Rooms

Frequency. Daily for offices, at least every other day for lockerrooms and rest rooms.

Procedure

1. Cover electrical connections with plastic sheeting.
2. Clean areas with a foam or high-pressure unit (or scrub and mop).
3. No more than 20 minutes after applying the cleaning compound, rinse with water at 50 to 55°C (122–131°F).
4. If these steps do not clean dirty areas or if drains are not present, hand-scrub with scouring pads.

TROUBLESHOOTING TIPS

Discolored floors: Lighten darkened concrete floors using a bleach solution (soak for at least 30 minutes). Finish cleaning the floor using a mechanical scrubber.

White film buildup on equipment: Too much cleaning compound, poor rinsing, or hard water cause this film. Do not use too much cleaning compound, rinse thoroughly, and use soft water.

Conveyor wheels freezing: The cleaning water is probably too hot. Wheels lose lubricant when the temperature reaches 90°C (194°F). The cleaning solutions should not be hotter than 55°C (131°F). Relubricate the conveyor wheels (follow the manufacturer's directions).

Sewer lines plugged: Clean sediment bowls daily, and do not flush floor sweepings into sewer pipes.

Yellow protein buildup on equipment: The water may be too hot. Brush away the organic material that builds up every day. If soil has been cooked onto equipment for a long time, rub with steel wool to remove the soil.

Do not spray: liver slicers, cube steak machines, electronic scales, patty machines, electrical outlets, motors, or equipment with open connections, wrapping film or containers, and wrapping units. Cover all outlets with polyethylene bags, and cover electrical equipment with waterproof drop cloths. Follow the manufacturer's recommendations when cleaning this equipment.

SUMMARY

- A good cleaning system can cut labor costs in meat and poultry plants by as much as 50%.
- *L. monocytogenes* is a common pathogen in meat- and poultry-processing plants.
- The optimal cleaning system depends on the type of soil and the type of equipment. High-pressure, low-volume cleaning equipment works best for removing heavy organic soil, especially in areas that are difficult to reach. But foams, slurries, and gels are quicker and easier to apply. CIP systems are expensive and do not work well in meat and poultry plants, except in large storage containers.
- HACCP is a good way to make sure that meat and poultry products are safe to eat.

- Operators of meat and poultry plants use acid cleaning compounds to remove mineral deposits and alkaline cleaning compounds to remove organic soils. Chlorine compounds are the best and cheapest sanitizers. But iodine compounds are less corrosive and irritating, and quaternary ammonium sanitizers have a longer-lasting effect.

- Operators should have written cleaning procedures for each area of the plant.

BIBLIOGRAPHY

Daun, H., Solberg, M., Franke, W., and Gilbert, S. 1971. Effect of oxygen-enriched atmospheres on storage quality of packaged fresh meat. *J. Food Sci.* 36:1001.

Douglas, G. S. 1987. Listeria and food processing. *Technics/Topics* 11:2.

Doyle, M. P. 1987. Low-temperature bacterial pathogens. *Proc. Meat Ind. Res. Conf.*, p. 51. American Meat Institute, Washington, D.C.

Ledward, D. A. 1970. Metmyoglobin formation in beef stored in carbon dioxide enriched and oxygen depleted atmospheres. *J. Food Sci.* 35:33.

Marriott, N. G. 1994. *Principles of Food Sanitation*, 3d ed. Chapman & Hall, New York.

Marriott, N. G. 1990. *Meat Sanitation Guide II*. American Association of Meat Processors and Virginia Polytechnic Institute and State University, Blacksburg.

Price, J. F., and Schweigert, B. S. 1987. *The Science of Meat and Meat Products*, 3d ed. Food & Nutrition Press, Westport, Conn.

Roback, D. L., and Costilow, R. M. 1961. Role of bacteria in the oxidation of myoglobin. *Appl. Microbiol.* 9:529.

Shapton, D. A., and Shapton, N. F., eds. 1991. Buildings. In *Principles and Practices for the Safe Processing of Foods*, p. 37. Butterworth-Heinemann, Oxford.

Smulders, F. J. M. 1987. Microbial contamination and decontamination. *Proc. Meat Ind. Res. Conf.*, p. 29. American Meat Institute, Washington, D.C.

Solberg, M. 1968. Factors affecting fresh meat color. *Proc. Meat Ind. Res. Conf.*, p. 32. American Meat Institute Foundation, Chicago.

STUDY QUESTIONS

1. Why doesn't refrigeration control *L. monocytogenes?*
2. List six steps to control *L. monocytogenes.*
3. Give five reasons why good sanitation is important in meat- and poultry-processing facilities.
4. What are the basic steps of a cleaning procedure?
5. Give three examples of CCPs.
6. Why are written sanitation procedures important?

TO FIND OUT MORE ABOUT MEAT AND POULTRY SANITATION

1. Call the United States Department of Agriculture (USDA) Meat & Poultry Hotline, (1-800) 535-4555 for answers to questions about handling meat and poultry, how to tell if it's safe to eat, and how to read meat and poultry labels. The hotline is staffed from 8 a.m.

to 5 p.m. (EST), Monday to Friday. You can also write to: The Meat and Poultry Hotline, USDA-FSIS, Room 1163-S, Washington, DC 20250.

2. Call the toll-free number on a package of processed meats, and ask if the manufacturer has information on food safety of their products.

3. Contact the National Cattlemen's Beef Association, (312) 467-5520, 444 North Michigan Avenue, Chicago, IL 60611, and ask for information on food safety and sanitation for red-meat products.

CHOOSING THE RIGHT CRITICAL CONTROL POINT AND THE RIGHT CRITICAL LIMIT

For HACCP to work, it is important to find out what the critical control points (CCPs) are, i.e., where the meat becomes most contaminated during processing. Each CCP then needs a critical limit, i.e., whatever is needed to keep the contamination low enough to be acceptable at that point.

This study found that inspecting lamb carcasses is not a good way to tell whether a carcass is heavily contaminated with microbes. Carcasses that had visible pieces of wool, skin, feces, or other intestinal matter were not always carrying more microbes than carcasses that looked clean. But carcasses from animals that had long wool or that had been washed before slaughter carried more microbes than animals with short wool or that had not been washed. Therefore, checking the length of the wool on the lambs when they arrive at the slaughterhouse and not prewashing the lambs are more important than checking carcasses on the line for obvious contamination.

This study shows that sanitarians need to do a careful hazard analysis and use microbiological testing when they set up the HACCP system, to make sure that the things they decide to monitor are really the most important points in the process where the food product could be contaminated.

Source: Biss, M.E., and Hathaway, S.C. 1995. Microbiological and visible contamination of lamb carcasses according to preslaughter presentation status: implications for HACCP. *J. Food Protection.* 58:776.

CHAPTER 13

Seafood Sanitation

ABOUT THIS CHAPTER

In this chapter you will learn:

1. How seafood can be contaminated
2. How the design and construction of seafood-processing plants can cut down on seafood contamination
3. How to clean and sanitize seafood-processing plants
4. How seafood processors can recover or recycle by-products
5. About inspection programs and HACCP in the seafood industry

INTRODUCTION

To keep the facility safe and hygienic, seafood processors need to know about:

- Microorganisms that cause spoilage and foodborne illness
- Types of soil they will need to remove
- Good cleaning compounds and sanitizers
- Types of cleaning equipment they can use
- Good cleaning procedures

Processors need to know about federal, state, and local public-health regulations. By law, seafood does not have to be continually inspected during processing. But recently, the possible risks associated with eating fish and shellfish have received a lot of publicity. Several bills have been introduced in both the U.S. Senate and House of Representatives calling for some type of mandatory seafood inspection to protect consumers from health hazards, economic fraud, and unacceptable or unsanitary manufacturing practices.

Regulations are not the only reason why seafood processors should use strict sanitary methods. Another important reason is that consumers are more and more aware that food, including fish and seafood, should be nutritious, wholesome, and processed under sanitary conditions.

Therefore, sanitation programs are essential so that seafood processors can provide consumers with high-quality, wholesome food. These guidelines cover the structure and work practices in new, old, expanded, and renovated plants. Every step from harvesting to eating must be controlled to make sure that only wholesome seafood products reach the consumer.

SOURCES OF SEAFOOD CONTAMINATION

The environment around a seafood plant can contaminate the plant and its products. The processing equipment, containers, and work surfaces can also contaminate seafood. A good sanitation program reduces the amount of contamination and checks that the program is working.

Seafood involves many different types of sea animals, and some types are more likely to be contaminated than others. Mollusks (e.g., mussels) are more likely to be contaminated than crustaceans (e.g., shrimp) or fin fish (e.g., trout, salmon, snapper). This is because mollusks feed by filtering seawater, so they can easily filter out contaminants as well as food. The raw product can be contaminated, especially if it is not harvested properly and the boat or truck that carries it is not sanitary. Delaying refrigeration after harvesting and other handling errors between harvesting and processing can cause the seafood to decompose (develop-off-conditions) and allow microbes to grow rapidly in the food.

The day after harvesting, the quality and safety of seafood are usually good if:

1. Chilling begins immediately after harvesting.
2. Chilling reduces the temperature of the product to 10°C (50°F) within 4 hours.
3. Chilling continues to approximately 1°C (34°F).

If a processor stores seafood at 27°C (81°F) or higher for 4 hours and then chills it to 1°C (34°F), the fish will be safe to eat for only 12 hours.

Workers can contaminate seafood, especially through poor hygiene. Also, processing equipment, boxes, belts, tools, walls, floors, utensils, supplies, and pests can contaminate seafood. The most serious problem is when microbes contaminate ready-to-eat foods. Therefore, workers must carefully clean and sanitize all equipment.

SANITATION MANAGEMENT

A seafood sanitation program must include proper sanitation practices and adequate staffing.

Staff

Seafood facilities need a well-qualified sanitarian. The seafood plant manager is responsible for making sure that there is a good sanitation program. Products must be wholesome, sanitation employees must be trained to keep the plant clean, and employees should know about seafood products and proper sanitation. Employees with infectious illnesses should not work in processing areas, even during cleaning. (See Chap. 4 for more information about employee health.)

Seafood-processing plants should have at least one employee responsible for inspecting all equipment and processing areas for hygienic conditions every day. Workers should reclean and sanitize anything that is not sanitary before they begin production.

The Cleaning Schedule

The plant must have a step-by-step cleaning schedule. Each area of the plant should have its own schedule, and supervisors should follow it rigidly. Workers should clean equipment that runs continuously (such as conveyors, filleting machines, batter and breading machines, cookers, and tunnel freezers) at the end of each production shift. If the area is not refrigerated, workers should clean batter machines and other equipment that holds milk or egg products every 4 hours. The worker should drain the batter, flush the batter reservoir with clean water, and apply a sanitizer. At the end of the production shift, workers should take the equipment apart and clean and sanitize each part. Workers should store machine parts and portable equipment off the floor in a clean area away from splash water, dust, and other types of contamination.

Workers should take the following steps when cleaning seafood plants:

1. Cover electrical equipment with polyethylene sheeting.
2. Remove large debris and put it in trash containers.
3. Manually or mechanically remove soil from the walls and floors by scraping or brushing, or by hosing with mechanized cleaning equipment. Start at the top of equipment and walls, and work down toward the floor drains or exit.
4. Take the equipment apart.
5. Prerinse with water at 40°C (104°F) or lower to wet the surfaces, and remove large and water-soluble debris. The temperature is important. A higher temperature can bake soil onto surfaces.
6. Apply a cleaning compound that works on organic soil (usually an alkaline cleaner) using portable or centralized high-pressure, low-volume or foam equipment. The cleaning solution should not be hotter than 55°C (131°F). Cleaning compounds such as a general-purpose cleaner or a chlorinated alkaline detergent usually work well. Use more than one cleaner to remove the different types of soil (See Chap. 5 for more information about cleaning compounds and Chap. 7 for more information about cleaning systems.)
7. Leave the cleaning compound for about 15 minutes to work on the soil, and rinse the equipment and area with water at 55 to 60°C (131–140°F). Hotter water removes fats, oils, and inorganic materials well but also increases energy costs and leaves more condensation on the equipment, walls, and ceilings. The cleaning compound emulsifies fats and oils at cooler temperatures.
8. Inspect the equipment and the facility, and reclean if necessary.

9. Make sure that the plant is microbially clean by using a sanitizer. Chlorine compounds are the most economical and most widely used, but are not the only method (see Chap. 6). Table 13.1 shows the recommended concentrations for various sanitizers. Sanitizers work best when staff use them in a portable sprayer in small operations or a centralized spraying or fogging system in large-volume operations. Chapter 7 also discusses equipment for applying sanitizers.

10. Avoid contaminating the equipment and the area during maintenance and setup. Require maintenance workers to carry a sanitizer and to use it where they have worked.

Further information about cleaning compounds, sanitizers, and cleaning equipment is provided in Chapters 5, 6, and 7.

TABLE 13.1. Recommended Sanitizer Concentrations for Various Areas

	Concentration, ppm		
Application	Available Chlorine	Available Iodine	Quaternary Ammonium Compounds
Washwater	2–10	Not recommended	Not recommended
Hand-dip	Not recommended	8–12	150
Clean, smooth surfaces (rest rooms and glassware)	50–100	10–35	Not recommended
Equipment and utensils	300	12–20	200
Rough surfaces (worn tables, concrete floors, and walls)	1,000–5,000	125–200	500–800

PLANT CONSTRUCTION

A well-designed plant can help make the sanitation program work better and more efficiently. But even a well-designed plant cannot stop microbial infestation or other contamination unless the facility also has a good maintenance and sanitation program. In a hygienic operation, the employer or management team insist on good housekeeping and are constantly watching out for sanitation problems in all buildings, areas, equipment, employees, and ingredients.

Site Requirements

Many seafood-processing sites are conveniently close to water. Engineering management should make sure that storm surges will not damage the facility or contaminate food products. Designers should look closely at receiving areas where raw materials arrive by boat or truck.

The site should be clean and attractive. Cleanliness and neatness help the firm's public image, and good first impressions are important to inspectors and the public. The condition of the outside of the plant often suggests the plant's hygiene standards inside. The Food and Drug Administration reports that areas not properly drained may contaminate food products through seepage and by providing a place for microorganisms and insects to grow. Very dusty roads, yards, or parking lots can contaminate areas where food is exposed. Also untidy refuse, litter, equipment, and uncut weeds or grass around the plant buildings may be a good place for rodents, insects, and other pests to breed.

The site should have a way to dispose of seafood wastes. Solids, liquids, vapors, and odors that come from the plant are not good signs and can lead to legal action by regulators or concerned citizens. Waste-disposal facilities must meet federal, state, and local requirements. It is best to get clearance or approval before building the facility or system to avoid expensive and time-consuming changes later on.

The site must also have enough drinkable water for the plant's needs. If the water comes from wells, sanitarians should analyze the water for minerals and microbes to make sure it meets regulations. The plant also needs a plan for its wastewater.

Construction

Builders must use materials that are easy to clean, resist corrosion, do not deteriorate in other ways, and do not absorb water. Air or mesh screens should cover all openings to keep insects, rodents, birds, and other pests out. The facility should be large enough to be well-organized, tidy, and sanitary. The sanitary features of various parts of the building are discussed here.

Floors. Floors should be made of hard, nonporous material, such as waterproof concrete or tile. The material should be durable and have an even surface so that it does not hold debris. But the surface should not be so smooth that it is slippery and dangerous. A rough finish or use of abrasive particles in the surface can cut down on accidents. Water-based acrylic epoxy resin is a popular, durable, nonabsorbent, and easy-to-clean surface. Acid brick floors are also good and durable but are very expensive.

Floor drains. The processing area needs a floor drain for each 37 m^2 (44 yds^2) of floor space. Floors in the processing areas should slope down to a drainage outlet (slope of 2%). The slope must be uniform, without dead spots that trap water and debris. All drains should have traps. Drainage lines should be at least 10 cm (4 in) wide inside and should be made of cast iron, steel, or polyvinyl chloride tubing. Designers need to check state and local codes to make sure that they allow these materials. Drainage lines should have vents to the outside air to cut down on odors and contamination. All vents should have screens so pests cannot use them to get into the plant. Drainage lines from toilets should connect right into the sewage system, not into other drainage lines, so that production areas cannot be contaminated by backup from the toilet areas.

Ceilings. Ceilings should be at least 3 m (10 ft) high in work areas and made of non-porous material, such as portland-cement plaster, with joints sealed by flexible sealing compound. A false ceiling can prevent debris from overhead pipes, machinery, and beams from falling onto food products or ingredients.

Walls and windows. Walls should be smooth and flat. They should be made of glazed tile, glazed brick, smooth-surfaced portland-cement plaster, or other nonabsorbent and nontoxic material. Concrete walls are good if they have a smooth finish. Painting is discouraged, but a nontoxic paint that is not light colored and not lead based can be used. Window sills should slant at a 45° angle so that debris cannot settle on them.

Entrances. Entrances should be made of materials that do not rust and have tightly soldered or welded seams. Outside entrances should have double-entry screened doors, and doors in the processing areas should have air curtains.

Processing Equipment

Processing equipment should have a lasting, smooth finish and be easy to clean. Surfaces should not have pits, cracks, and scale. The equipment design should stop lubricants, dust, and other debris from contaminating food products. Food processors should install and maintain the equipment with enough room around it for cleaning the outside surfaces and surrounding areas. Food processors should follow the equipment standards of the Association of Food and Drug Officials of the United States (AFDOUS) (1962).

Metal equipment should be made of stainless steel to protect seafood and other food products. Seafood products, cleaning compounds, and salt water can corrode galvanized metal, so it should not be used for processing equipment. But galvanized metal can be an economical material for handling waste materials. Any galvanized metal should be smooth and have a high-quality finish.

Food processors should use cutting boards made of a hard, nonporous, moisture-resistant synthetic material such as polyethylene. They should not use wooden boards because they absorb moisture and are difficult to clean. Cutting boards should be easy to remove for cleaning and should be kept smooth; should be abrasion and heat resistant, shatterproof, and nontoxic; and should not contain material that will contaminate seafoods.

Conveyor belts should be made of moisture-resistant material (such as nylon or stainless steel) that is easy to clean. Conveyors should not have hard-to-reach corners that collect debris. The equipment should come apart easily for cleaning. Drive belts and pulleys should have protective guard shields that are easy to remove during cleaning. Motors and oiled bearings should be placed where oil and grease will not contaminate the food product.

Equipment that does not move should not be closer than 0.3 m (1 ft) away from walls and ceilings to allow access for cleaning. Equipment should either be 0.3 m (1 ft) above the floor or have a watertight seal with the floor. All wastewater should leave through storage tanks so that it can be constantly fed into the drainage system without flowing over the floor.

RECOVERING BY-PRODUCTS

Waste management, including recycling seafood waste products, is becoming more and more important. A good recovery system can make the operation more hygienic, as well as save money. Many food processors now recycle or reduce their liquid waste.
   New ways to conserve water include:

- Use of uncontaminated wastewater from one area of a food-processing operation (e.g., water from the final rinse in a cleaning cycle) in other areas that do not need drinkable water
- Use of closed water systems in food-processing operations in which all water used in processing is continuously filtered to remove solids
- Use of dry conveyors to transport solids, instead of water

VOLUNTARY INSPECTION PROGRAMS

The seafood industry has a voluntary seafood inspection program conducted by the U.S. Department of Commerce, National Marine Fisheries Service. This systematic sanitation inspection program helps the industry begin and maintain sanitary operations. The program has helped to increase the amount of wholesome seafood that is produced and eaten.

Hazard Analysis Critical Control Point

In 1968, the seafood industry began testing a HACCP program for breaded and cooked shrimp at nine plants in different areas of the United States. Use of HACCP is increasing throughout the seafood industry. Boxed text at the end of this chapter gives more information about how seafood processors can use HACCP programs.
   Seafood inspection is not required by law. If an inspection law is passed, it is not yet known which federal organization will oversee the program. Scientists and industry representatives disagree as to whether seafood inspection should be mandatory and if it should be under the U.S. Food and Drug Administration, U.S. Department of Agriculture, National Marine Fisheries Service, or a combination of these organizations. A seafood inspection bill will probably be passed soon and will be based on HACCP principles.

SUMMARY

- Hygienic plant design helps seafood processors produce wholesome seafood products.
- The location of the seafood plant affects sanitation inside the facility.

- The design of the plant and the materials used to construct the plant and equipment are also important to the sanitation program.
- A hygienic operation depends on having enough staff for cleaning and an organized, step-by-step cleaning schedule.
- The sanitation program should use the best cleaning compounds, cleaning equipment, and sanitizers.
- Recovering by-products, using recommendations from regulatory agencies, and participating in voluntary inspection programs improves the sanitation program.

BIBLIOGRAPHY

AFDOUS (Association of Food and Drug Officials of the United States). 1962. *J. Assoc. Food Drug Off.* 26:39.

Jowitt, R. 1980. *Hygienic Design and Operation of Food Plant.* AVI Publishing Co., Westport, Conn.

Lane, J. P. 1979. Sanitation recommendations for fresh and frozen fish plants. In *Sanitation Notebook for the Seafood Industry,* p. 11–9, G. J. Flick, Jr., et al., eds. Department of Food Science and Technology, Virginia Polytechnic Institute and State University, Blacksburg.

Marriott, N. G. 1994. *Principles of Food Sanitation,* 3d ed. Chapman & Hall, New York.

Marriott, N. G. 1980. *Meat Sanitation Guide II.* American Association of Meat Processors and Virginia Polytechnic Institute and State University, Blacksburg.

Nardi, G. C. 1992. Seafood safety and consumer confidence. *Food Protection Inside Report* 8:2A.

Neal, C. L. 1979. Sanitary standards under the Department of Commerce Voluntary Seafood Inspection Program. In *Sanitation Notebook for the Seafood Industry,* p. IV-12, G. J. Flick, Jr., et al., eds. Department of Food Science and Technology, Virginia Polytechnic Institute and State University, Blacksburg.

Nickelson, R. II. 1979. Food contact surfaces—indices of sanitation. In *Sanitation Notebook for the Seafood Industry,* p. II—50, G. J. Flick, Jr. et al., eds. Department of Food Science and Technology, Virginia Polytechnic Institute and State University, Blacksburg.

Perkins, B. E., and Bough, W. A. 1979. Seafood industry sanitation through water conservation and by-product recovery. In *Sanitation Notebook for the Seafood Industry,* p. III-17, G. J. Flick, Jr., et al., eds. Department of Food Science and Technology, Virginia Polytechnic Institute and State University, Blacksburg.

STUDY QUESTIONS

1. How should seafood be chilled to make sure that it is safe to eat the day after it is harvested?
2. What are the basic cleaning steps in a seafood-processing facility?
3. Why should the outside of seafood-processing plants be clean and tidy?
4. Describe two ways that seafood-processing plants can reduce the amount of water they use.
5. Why is a HACCP program better than testing end products?

To Find Out More About Seafood Sanitation

1. Contact the Seafood Hotline, a service of the U.S. Food and Drug Administration Center for Food Safety and Applied Nutrition, (1-800) FDA-4010, and ask for information about seafood safety. Public affairs specialists are available noon to 4:00 p.m. EST, Monday to Friday.

2. Contact the Food Marketing Institute, (202) 429-8298, 800 Connecticut Ave., NW, Suite 500, Washington, DC 20006-2701. Ask for information about their HACCP information and materials for the seafood industry.

3. Contact the National Fisheries Institute, (703) 524-8881, 1525 Wilson Blvd. Suite 500, Arlington, VA 22209. Ask for information on seafood inspection, HACCP, foodborne illness from seafood, and seafood safety guidelines.

HACCP in the Seafood Industry

Traditionally, food processors have inspected and tested final food products to prevent foodborne illness. But this system does not control the risks. Foods rarely contain dangerous pathogens, and tests for microbes in foods are not very accurate or precise. Therefore, it is easy to miss a contaminated food product. This system is expensive and can give food processors a false sense of security.

A strategy to prevent foodborne illness by analyzing the entire food production process costs less and is more likely to ensure safety. Hazard Analysis Critical Control Point (HACCP) studies all factors that can lead to contamination of foods or growth of microorganisms during food processing. At the moment, it is considered the best system available for making sure that food products are safe. Chapter 10 describes HACCP in detail.

The way food processors apply the HACCP concept is unique for each food and factory. Each case needs a detailed study to identify the hazards and the critical control points (CCPs). But some principles apply to seafoods with similar types of microbes or types of processing.

Researchers have classified seafoods into hazard categories in order of decreasing safety risk (A is most risky, G is least risky).

(A) Mollusks, including fresh and frozen mussels, clams, and oysters. Often eaten with no additional cooking.

(B) Lightly preserved fish products, including salted, marinated, and cold smoked fish. Eaten without cooking.

(C) Heat-processed (pasteurized, cooked, hot smoked) fish products and crustaceans (including precooked, breaded fillets). Some products eaten with no additional cooking.

(D) Heat-processed (sterilized, packed in sealed containers). Often eaten with no additional cooking.

(E) Semipreserved fish, including salted or marinated fish and caviar. Eaten without cooking.

(F) Dried, dry-salted, and smoked dried fish. Usually eaten after cooking.

(G) Fresh and frozen fish and crustaceans. Usually eaten after cooking.

The hazard analysis of these products is fairly simple. The live seafood are caught in the sea or in freshwater, handled, and—in most cases—processed without additives or chemical preservatives. Most are chilled or frozen to preserve them during shipping. Most fish and shellfish are cooked before eating, although some are eaten raw (e.g., Japanese food). Most food-poisoning outbreaks after fish were eaten—fish that was cooked right before it was eaten—have been caused by such heat-stable toxins as biotoxins and histamine. Pathogenic microbes in fish need temperatures above 5°C (40°F) to grow. So when seafood is refrigerated, food spoilage bacteria are likely to cause the fish to go off long before the pathogens produce toxins or multiply to large numbers of pathogens. But if fish are kept at temperatures above 5°C (40°F), histamine- or biotoxin-producing bacteria have a chance to produce enough toxin or histamine to cause poisoning.

Fish caught in some areas may be infected with parasites that can also infect humans. The disease the parasites cause (e.g., anisakisis) is not fatal, and the parasites are destroyed if the fish is cooked, but parasites could be a problem if the fish is eaten raw. Freezing also kills the parasites if the temperature is below −20°C (−4°F) for 24 hours.

CCPs can be found for most of these hazards. The fishing industry cannot and does not need to control the bacteria in the seafood's natural environment. But regulators can limit animal and human bacteria that contaminate live fish by controlling fishing in polluted areas. They can also monitor fishing areas for the presence of parasites and biotoxins. Processors can control growth of bacteria by strict temperature control. A temperature below 3°C (37°F) is a CCP for controlling pathogens and histamine-production. Monitoring temperatures during handling and processing is a vital part of HACCP in the seafood industry.

When the HACCP system begins, it is important to check quite often that the system is working properly. This is done using microbiological, chemical, and sensory tests of the end product and at various stages of production. Once the system is established, processors only need to test occasionally.

Source: Huss, H. H. 1992. Development and use of the HACCP concept in fish processing. *Int. J. Food Microbiol.* 15:33–44.

CHAPTER 14

Fruit and Vegetable Processing and Product Sanitation

ABOUT THIS CHAPTER

In this chapter you will learn:

1. How to reduce contamination of fruit and vegetable products from raw products, soil, air, water, and pests
2. How good housekeeping and waste-disposal practices can help keep fruit and vegetable products wholesome
3. How the design of fruit- and vegetable-processing plants can affect hygiene
4. Methods for cleaning and sanitizing in fruit- and vegetable-processing plants
5. How sanitarians evaluate hygiene in fruit- and vegetable-processing plants

INTRODUCTION

Like other food operations, facilities that process fruits and vegetables should have a sanitation program that uses correct cleaning compounds, sanitizers, and cleaning procedures. Management must oversee and evaluate the sanitation program. The ultimate goal is to produce food products that are sanitary and wholesome.

REDUCING CONTAMINATION

Effective sanitation prevents microorganisms that cause spoilage and food poisoning from getting into fruits and vegetables during production, processing, storage, and distribution. It is important to remember that raw fruits and vegetables can have microorganisms on the surface or inside when they arrive at the plant. These microorganisms can contaminate the processing plant.

Federal laws require that processed foods be free of pathogenic microorganisms if they are shipped between states. For commercially canned foods, the normal sterilization process destroys all pathogenic bacteria in the container. Washing and peeling of fruits and vegetables also help remove microorganisms. Therefore, properly canned and frozen fruits and vegetables should be wholesome.

Contamination from Soil

The soil in which vegetables grow contains heat-resistant bacteria. If processors do not wash vegetables thoroughly before they are canned, these bacteria can cause a sour taste and aroma and other types of spoilage. The number and types of microbes are affected by wind, humidity, sunlight, temperature, domestic and wild animals, irrigation water, bird droppings, harvesting equipment, and workers. Many pathogens reach fruits and vegetables during irrigation shortly before harvesting and before the sun dehydrates and destroys the pathogens. Others reach them during harvesting and shipping.

Contamination from Air

Contaminated air transports microorganisms (including pathogens) and pollutants. Air filters can prevent unclean air from entering the food-processing facility.

Contamination by Pests

Some pests invade fruits and vegetables while they are forming on the tree or vine. Pests spread viruses, spoilage bacteria, and pathogens, and also cause physical damage to the fruit or vegetable. The pests are often inactive while the fruit or vegetable is forming because their skins keep the pests from getting inside, and there is very little moisture (low A_w) on the surface. As the fruit or vegetable matures, the structure changes and the pests can grow. When the protective skins of fruits and vegetables are broken by a bruise, mechanical injury, or attack by insects, microorganisms can get inside.

The pollinating fig wasp is a pest that carries microbes that live and grow in the fig while it ripens. Some microbes do not cause spoilage themselves, but they attract other organisms, such as fruit flies, which carry spoilage yeasts and bacteria.

Pests also enter fruits and vegetables during transport. Produce is especially vulnerable at this time if it has cuts or bruises from harvesting.

Processors should not use recirculated water for washing fruits and vegetables because microorganisms build up in the washwater. Chlorinating the washwater does not help because bacterial spores are resistant to chlorine.

Contamination During Processing

Traditional fruit and vegetable canning involved pouring food into containers (metal, glass, or plastic), sealing, and heat treatment. The heat treatment is called *terminal sterilization*. The time and temperature are carefully planned to kill *Clostridium botu-*

linum spores and most heat-resistant spores of other spoilage organisms. Canned foods have what is called "commercial sterility" (i.e., not quite sterile but enough bacteria and spores are destroyed for the food to be safe). In aseptic packaging or aseptic canning, the food and containers are commercially sterilized separately. The food is cooled, and the containers are filled and sealed in aseptic conditions.

In terminal sterilization, the sealed containers are heated (kill step). Because the containers are well made and durable, conventional canning is a very safe way to process food. This technology can also use HACCP to ensure safety.

Aseptic packaging is a relatively new technology. Therefore, it is important to find the best test methods to make sure that the package stays intact and sterile during distribution.

SANITATION MANAGEMENT

As in other food plants, managers are legally and morally responsible for providing consumers with wholesome products. An effective sanitation program keeps the processing environment clean.

Housekeeping

Housekeeping means being organized and tidy. Supervisors should carefully arrange supplies, materials, and clothing to keep the operation tidy, thus less likely to be contaminated and easier to clean.

The sanitarian should be responsible for housekeeping, but all production, maintenance, and sanitation workers need to work together. Employees must keep trash containers, tools, supplies, and personal belongings in their proper places. Trash is more likely to be thrown away immediately if trash containers are in convenient places.

Insects, rodents, and birds can contaminate foods. When sanitation is good, pests cannot find food and shelter in the plant. Use of air and mesh screens; filling in holes, cracks, and crevices; and other hygienic designs help keep pests out of the plant. Regular pest inspections are also important. (Pest control is discussed in Chap. 9.)

Waste Disposal

Waste is easier to handle and by-products are easier to salvage when solid and liquid wastes are separated. Waste disposal is discussed in Chapter 7. Some food-processing plants use waste by-products. The citrus juice industry is very efficient at salvaging waste and keeping waste-disposal costs down. More than 99% of the raw fruit is made into juices, concentrates, or dried cattle feed.

Water Supply

A plentiful supply of clean water is needed for a wholesome product and a clean plant. Water is not just a cleaning medium; it is also used to heat and cool products, and as an ingredient.

Staff should check levels of bacteria and organic or inorganic impurities in the water every day. The number of bacteria determines whether the water is safe to use as a food ingredient or to clean any surface that could touch or contaminate the food. Organic and inorganic impurities affect how well the water will wash the food, surfaces, and equipment.

PLANT CONSTRUCTION

Even a well-designed processing plant does not stop microbes from reaching the food unless each area and each piece of equipment are easy to clean. Proper instructions should be posted for each cleaning operation. If a processing plant is being newly built, expanded, or renovated, all professional staff (mechanical engineers, industrial engineers, food chemists, microbiologists, sanitarians, and operations personnel) should study the functional layouts, mechanical layouts, plumbing layouts, and equipment and construction specifications. Input from each of these people helps make sure that nothing, including sanitation, is overlooked.

Today, new and expanded fruit- and vegetable-processing plants must use hygienic designs because they process such a high volume. High-volume plants are more mechanized. Mechanized startup, shutdown, cleaning, and sanitizing mean less chance for human error but more chance of missing cleaning errors. Highly mechanized plants use more cleaning-in-place (CIP) systems, and less manual cleaning, and fewer visual inspections. However, CIP equipment is still rare in fruit- and vegetable-processing plants, except those that manufacture juices.

High-volume processing plants have longer production periods than lower-volume plants. More microbes build up in the plant because they have a longer time to accumulate and more food goes through the system. Saturation devices can (1) sense when microbes reach an unsafe level, (2) stop production, and (3) trigger an automatic cleaning procedure. These devices should probably be set so that they work only if microbe levels are very high, for example, 150% of normal levels.

Sanitary design features keep downtime for cleaning and sterilizing to a minimum. To be cost effective, equipment and facilities need to operate as much as possible and produce as little sewage as possible.

Processors can now use mechanized and automatic systems to clean equipment that used to have to be cleaned by hand. Before CIP was developed, machines and storage equipment were taken apart at the end of every production day and hand cleaned. CIP used to be controlled using a control panel with push buttons. Now panels have computer-controlled timers that automatically start and stop cleaning, rinsing, and sanitizing steps. (CIP is discussed in Chap. 7.)

Hygienic plant designs have no crevices (narrow and deep cracks or openings) or pockets (large cracks and openings) in the buildings and equipment. Crevices are harder to clean than pockets because they are harder to reach.

Principles of Hygienic Design

Newly constructed or remodeled fruit- or vegetable-processing plants should meet at least the following hygienic design standards:

- Equipment should be designed so that workers can take apart all surfaces that touch the food for manual cleaning or CIP.
- Outside surfaces should not allow soil, pests, and microorganisms to collect on the equipment or in the production area (walls, floors, ceilings, hanging supports, etc.).
- The equipment should protect food from contamination.
- All surfaces that touch food should be inert, so that they do not react with the food and are not absorbed by the food.
- All surfaces that touch the food should be smooth and nonporous, so that food particles, insect eggs, and microorganisms cannot collect in tiny cracks or scratches on the surface.
- The inside of the equipment should have as few crevices and pockets as possible.

The inside and outside of the plant should be sanitary. The following should be avoided so that debris cannot collect:

- Ledges and dirt traps
- Bolts, screws, and rivets that stick out
- Recessed corners, uneven surfaces, and hollows
- Sharp edges
- Unfilled edges

The plant should have double-door construction, heavy-duty weather strips, and self-closing door mechanisms to prevent pests from entering the plant.

It is important to make sure that the facility is designed and built properly, otherwise it may need expensive renovations. Also, the plant must be built carefully to make sure that areas and equipment are not contaminated during construction. The layout must be flexible, because the processing system may change and use new technology. The following points can help reduce contamination:

- The plant should have adequate storage space for raw materials and supplies. The area should be big enough to inspect raw materials before they are stored to make sure they are not carrying pests or dirt. Contaminated supplies should be separated from the rest and cleaned so that contamination does not spread. Cleaning and maintenance materials can taint raw materials if they share the same storage area.
- Finished food products should be stored separately from raw materials. If there is not enough space, the production area may be used to store the finished product. However, this can cause cross-contamination of raw materials while they are being processed.
- Open food production areas should not be blocked or cluttered, so that staff can clean and maintain equipment easily. Staff injuries and equipment damage are more common in messy work areas.
- Waste-removal routes should be short and direct, so that staff do not carry waste through open production areas. This is important because waste collection equipment is often unsanitary.

- Returned goods are often infested or partially decomposed. They must be stored away from raw material and production areas.
- The area around the building should be designed to keep pests out. Waste should be collected, treated, and incinerated away from the plant and away from air intake vents. The outside area should be easy to clean, well drained, and free of weeds, discarded equipment, and surplus supplies.
- Employees must have excellent personal hygiene (see Chap. 4).

CLEANING PROCESSING PLANTS

Cleaning equipment needs an efficient layout to save cleaning time. If possible, the cleaning layout should be designed when the facility is built, expanded, or renovated. It is much easier to install cleaning equipment at the same time as the processing equipment than to add it later. The soil found in fruit- and vegetable-processing plants is easy to remove using portable cleaning systems in small plants and a combination of CIP and centralized foam cleaning in large plants.

Hot-Water Wash

Water carries cleaning compounds and suspended soil. It removes sugars, other carbohydrates, and other compounds that dissolve in water. A hot-water (60–80°C, 140–176°F) wash requires very little cleaning equipment. But it uses a lot of labor and energy and causes condensation on equipment and surfaces. This cleaning technique does not remove heavy soil deposits very well.

High-Pressure, Low-Volume Cleaning

High-pressure spray cleaning works well for removing heavy soils in the fruit- and vegetable-processing industry. This method can clean difficult-to-reach areas with less labor, and the cleaning compounds work better at temperatures below 60°C (140°F). Water that is hotter than this can bake the soil onto the surface and speed up growth of microbes.

A high-pressure, low-volume system uses portable hydraulic units that can be easily moved throughout the plant. Portable equipment works well for conveyors and processing equipment where soaking is impractical and manual cleaning is difficult and time consuming. Centralized high-pressure, low-volume equipment uses the same principle as the portable unit but provides a higher output per man-hour of labor. Most suppliers of these systems can give customers technical assistance to make sure the equipment and cleaning compounds work properly.

Foam Cleaning

Portable foam cleaning is used a lot in fruit- and vegetable-processing plants because the foam is quick and easy to apply to ceilings, walls, piping, belts, and storage containers. The equipment is about the same size and costs about the same as portable high-pressure units.

Centralized foam cleaning uses the same technique as portable foam equipment and the equipment is installed at convenient places in the plant. The cleaning compound mixes with water and air automatically to form a foam.

Gel Cleaning

With this method, the cleaning compound is used as a gel rather than as a spray or foam. Gel works well for cleaning, canning, and packaging equipment because it clings to the surface.

Slurry Cleaning

This method is the same as foam cleaning, except that less air is mixed with the cleaning compound. A slurry is more liquid than foam and penetrates uneven surfaces in a canning plant better, but it does not cling as well as foam.

Combination Centralized High-Pressure and Foam Cleaning

This system is the same as centralized high pressure, except that the equipment can also apply foam. This method is more flexible because foam can be used for large surface areas, and high pressure can be used for belts, stainless-steel conveyors, and hard-to-reach areas in a canning plant.

Cleaning in Place (CIP)

With this closed system, a recirculating cleaning solution is sprayed into the equipment through nozzles and automatically cleans, rinses, and sanitizes equipment. CIP equipment is expensive and does not work well on heavy soil. But CIP cleaning is used in vacuum chambers, pumping and circulation lines, and large storage tanks. High-volume operations use CIP cleaning because it saves so much labor, which pays for the equipment more quickly. See Chapter 7 for more information about cleaning equipment.

CLEANERS AND SANITIZERS

Soil that is not removed during cleaning is contaminated with microorganisms. Soil can protect microorganisms so that they do not come in contact with chemical sanitizers. Also, soil dilutes chemical sanitizing solutions.

If the soil is light and the temperature of the cleaning solution is below 60°C (140°F), a one-step combined chemical sanitizer and cleaning compound may be used. Combination cleaners (detergent-sanitizers) are often used for manual cleaning in smaller operations with a solution temperature below 60°C (140°F). If the cleaning solution is hotter than 80°C (176°F), the solution will destroy most pathogenic and spoilage microorganisms without using a chemical sanitizer. The best chemical sanitizers for fruit- and vegetable-processing plants are halogen compounds, quaternary ammonium compounds, and phenolic compounds.

Halogen Compounds

Chlorine and chlorine compounds are the best halogen sanitizers for food-processing equipment and containers and for water supplies. Calcium hypochlorite and sodium hypochlorite are often used in fruit- and vegetable-processing plants. Plain chlorine is cheaper, but calcium hypochlorite and sodium hypochlorite are easier to use in dilute solution. Hypochlorite solutions are sensitive to temperature changes, residual organic soil, and pH (acidity). These compounds work quickly and are cheaper than other halogens, but they tend to be more corrosive and irritating to the skin. Chapter 6 has more information about chlorine and iodine sanitizers.

Quaternary Ammonium Compounds

Quaternary ammonium compounds ("quats") work well against most bacteria and molds. Quats are stable as a dry powder, a concentrated paste, or in solution at room temperature. They are not affected by heat, dissolve in water, have no color or smell, do not corrode metals, and do not irritate skin. These compounds work better than other sanitizers when there are soil residues. They work best against microbes when the pH is slightly acid or is alkaline (pH ≥6.0). The quats do not work well against bacteria when they are combined with cleaning compounds or in hard water.

Phenolic Compounds

These compounds are often used in antifungal paints and antifungal protective coatings, rather than as sanitizers applied after cleaning. Phenolic compounds have little use in fruit and vegetable plants because they do not dissolve well in water.

CLEANING PROCEDURES

As with other types of food plant, one set of procedures will not work for every fruit- and vegetable-processing plant. Procedures will vary with the design, size, age, and condition of the plant. The guidelines that follow should be adapted for each operation.

Making Cleaning Easier

Cleaning is easier if employees take steps to reduce the amount of soil and clean or rinse equipment quickly after use.

1. Control heating of equipment so that soil does not burn on.
2. Rinse and wash equipment immediately after use so that soil does not dry on.

3. Replace faulty gaskets and seals so that they do not leak and splatter.
4. Handle food products and ingredients carefully so that they do not spill.
5. Keep all areas tidy.
6. If equipment breaks down, rinse equipment and cool to 35°C (95°F) to slow down growth of microbes.
7. During brief shutdowns, keep washers, dewatering screens, blanchers, and similar equipment running and cooled to 35°C (95°F) or below.

Preparing to Clean

1. Remove large debris from the area.
2. Take equipment apart as much as possible.
3. Cover electrical connections with plastic film.
4. Disconnect lines or open cutouts so that debris is not washed onto other equipment that has been cleaned.
5. Remove large chunks of waste from equipment using an air hose, broom, shovel, or other appropriate tool.

Processing Areas

Frequency. Daily.

Procedure
1. Prerinse all soiled surfaces with water at 55°C (131°F) to move dirt down from the ceilings and walls to the floor drains. Do not hose motors, outlets, and electrical cables directly.
2. Use portable or centralized foam cleaning equipment and a strongly acid cleaner. A centralized system works best for large plants; portable equipment is more practical for smaller plants. If the equipment is made of a metal other than stainless steel, a heavy-duty alkaline cleaning compound should be used, rather than an acid cleaner. If necessary, hand-brush soil deposits that stick after foam cleaning. The cleaning compound should reach all framework, table undersides, and other difficult-to-reach areas. The cleaning compound should soak for 10 to 20 minutes.
3. Rinse surfaces within 20 minutes after applying the cleaner to remove the residues. The area should be rinsed from top to bottom using water at 50–55°C (122–131°F).
4. Inspect all surfaces thoroughly, and do touch-up cleaning, if necessary.
5. Use centralized or portable sanitizing equipment and a chlorine compound sanitizer. (The concentration should be 100 ppm of chlorine.) Use the same method to sanitize water pipes used to recirculate washwater and to pump brines, syrups, and vegetables such as peas and corn. Frequently drain, clean, and sanitize water storage tanks so that microbes do not build up.
6. Thoroughly backwash and sanitize water filters and water softeners.

7. Remove scale from the surfaces of pipes and equipment, because scale can collect microorganisms.
8. Remove, clean, and replace drain covers.
9. Apply a white edible oil only to surfaces that may rust or corrode. Do not use oil anywhere else because it can collect microorganisms.
10. Require that maintenance workers carry a sanitizer and use it where they have worked.

Large processing plants can use a CIP system for cleaning piping, large storage tanks, and cookers. The CIP system can be used instead of steps 1, 2, 3, and 5.

Packaged Storage Areas

Frequency. At least once a week where processed products are stored, and more often in a high-volume operation; daily in areas where raw products are stored.

Procedure
1. Pick up large debris and place in trashcans.
2. Sweep and/or scrub the area. Use a mechanical sweeper or scrubber, if possible. Use the correct cleaning compounds for the mechanical sweeper or scrubber, and follow the vendor's directions.
3. Use a portable or centralized foam or slurry cleaning system with water at 50°C (122°F) to clean heavily soiled areas. Rinse as described for the processing areas.
4. Remove, clean, and replace drain covers.
5. Replace hoses and other equipment.
6. Wash and sanitize vegetable boxes each time they are emptied. Replace wooden husker and cutter bins with metal containers.

EVALUATION

The sanitarian must evaluate the sanitation program to see if all areas and equipment are clean and sanitary. Written records are important to show that the facility is meeting its sanitation goals.

Laboratory Tests

The sanitarian must know the type, characteristics, and sources of microorganisms found in the plant so that he or she knows which laboratory tests to use. Laboratory tests can help monitor how well the sanitation program is working. The sanitarian should try to reduce the total number of microorganisms on clean equipment and in processed food products. However, the total number of microbes is not always related to the number of spoilage or pathogenic microorganisms. The sanitarian needs to identify coliforms because they indicate contamination, and identify thermophiles and certain mesophiles because they may cause spoilage. Large numbers of spore-forming bacteria can shorten the shelf life of the food product and may cause food-borne illness.

Spot laboratory checks for microbial load can be used when visual inspections suggest a problem. The sanitarian should take samples from products and equipment at various stages during processing to identify trouble spots. Regular laboratory tests encourage workers to think about and use good sanitary practices.

SUMMARY

- A good sanitation program for fruit- and vegetable-processing facilities includes:

  Hygienic design of facilities and equipment

  Training of sanitation personnel

  Use of the right cleaning compounds and sanitizers

  Use of good cleaning procedures

  Good administration of the sanitation program

  Evaluation of the program using visual inspection and laboratory tests

- Contamination of raw materials, water, air, and supplies should be kept to a minimum.
- Hygienic design makes the equipment and facility easier to clean.
- Portable or centralized high-pressure or foam cleaning systems and CIP systems can reduce cleaning labor.
- Acid cleaning compounds work well in many facilities, if equipment and surfaces are made of durable material. Chlorine compound sanitizers work best and are most economical. Phenolic compounds can be added to paints and other protective coatings.
- The sanitarian should use visual inspection and laboratory tests to evaluate the sanitation program.

BIBLIOGRAPHY

Jowitt, R. 1980. *Hygienic Design and Operation of Food Plant.* AVI Publishing Co., Westport, Conn.

Lopez, A. 1987. *A Complete Course in Canning (Book I),* 12th ed. Canning Trade, Baltimore.

Marriott, N. G. 1994. *Principles of Food Sanitation,* 3d ed. Chapman & Hall, New York.

STUDY QUESTIONS

1. How are microorganisms destroyed during canning?
2. What are the three main sources of contamination for processed fruits and vegetables?

3. Why is it important to keep production areas tidy?

4. Why is it important to keep the outside of the building neat and clean?

5. What types of fruit- and vegetable-processing plants are likely to use CIP?

6. What is the difference between traditional canning and aseptic canning?

7. Why does the sanitarian need to know which microorganisms are likely to be present?

TO FIND OUT MORE ABOUT FRUIT AND VEGETABLE PROCESSING AND SANITATION

1. Contact the Produce Marketing Association, 700 Barksdale Plaza, Newark, DE 19711, (302) 738-7100. Ask for information about food safety and sanitation.

2. Contact the National Grocers Association, 1825 Samuel Morse Drive, Reston, VA 22090-5317, (703) 437-5300. Ask for information about food safety and sanitation.

RAW MATERIALS CAN BE CONTAMINATED!

Farmers need to be careful that food is not contaminated at the farm after it is harvested. *Cryptosporidium* is an intestinal parasite that causes diarrhea. It can be transmitted by contaminated water (waterborne) or from person to person (for example, in child care or hospital settings, where people who change diapers or help people use the toilet also serve food). But in October 1993, fresh-pressed apple cider from contaminated apples caused a rare case of foodborne cryptosporidiosis.

Two elementary schools and a high school in a rural farming community in central Maine held a 1-day agricultural fair. Local farmers provided agricultural demonstrations, a petting zoo of farm animals, a hay ride, a cider-pressing demonstration, and light refreshments. Eight days after the fair, principals of both elementary schools noticed that many children who had attended the fair were absent from classes.

Investigators collected evidence that showed this was a foodborne outbreak of cryptosporidiosis.

- Of the 759 people who went to the fair, 611 completed surveys.
- People who had diarrhea for at least 3 days totaled 160.
- Of those people who were ill, 56 provided stool samples for testing; 50 samples contained *Cryptosporidium*.
- Diarrhea started 10 hours to 13 days after the fair.
- Diarrhea lasted from 1 to 16 days.
- Of the 284 people who drank apple cider in the afternoon, 154 developed cryptosporidiosis.

- Of the 292 people who did not drink apple cider in the afternoon, only 6 developed cryptosporidiosis.
- *Cryptosporidium* was found in the apple cider, on the cider press, and in a stool specimen from a calf on the farm that supplied the apples.

The apples were probably contaminated by calf feces on the ground before or during harvest. The apples were probably not washed properly before they were pressed to make cider, so *Cryptosporidium* spread throughout the batch of cider.

Investigators warn farmers that farm animals should not graze in areas where fruits and vegetables are grown. Food processors should throw out any fruits and vegetables that have feces on them, and should wash fruits and vegetables thoroughly before they process them.

Source: Millard, P. S., Gensheimer K. F., Addiss, D. G., Sosin, D. M., Beckett, G. A., Houck-Jankoski, A., and Hudson, A. An outbreak of cryptosporidiosis from fresh-pressed apple cider. *J. Am. Med. Assoc.* 272(20): 1592.

CHAPTER 15

Beverage Plant Sanitation

ABOUT THIS CHAPTER

In this chapter you will learn:

1. How sanitation is important to producers of nonalcoholic beverages, beer, wine, and distilled spirits
2. Why beverage plants are easier to clean than food-processing plants
3. Why it is essential to make sure that the raw ingredients used in beverages are pure
4. How to clean and sanitize different types of beverage plant
5. Why beverage producers need good-quality water and how they obtain it
6. Why sanitation needs to be strictest in the bottling area

INTRODUCTION

Soils found in beverage plants contain mostly sugar and dissolve in water easily, so beverage plants are easier to clean than other food-processing plants. Cleaning and control of microbes are more difficult in breweries and wineries than in plants that produce nonalcoholic beverages. Therefore, most of this chapter will concentrate on these two areas.

FOCUS ON YEASTS

In beverage plants such as breweries, it is important to keep the useful microbes and remove those that cause spoilage and disease. Effective sanitation can keep microorganisms under control, but the plant will never be completely free of microorganisms. So if sanitation is poor, microorganisms soon contaminate beverages. Breweries are very similar to other manufacturing plants, although the most common pathogens are not usually a problem, mainly because of the raw materials, processing techniques, and characteristics of the final product (acidity, high alcohol concentration, and carbon dioxide content). It is important to make sure that the raw materials are kept sanitary throughout processing because there is no good way to detoxify contaminated beverages once they are made.

SANITATION MANAGEMENT

Toilet facilities must be adequate, kept clean, and located close to the bottling area and other production areas. Employees must wash their hands after using toilet facilities. Drinking fountains should have guards so that a person's mouth or nose cannot touch the water outlet.

Employee Practices

As with other food operations, sanitation is a team job. In beverage plants, employees should clean as they go. Regular cleaning helps keep the plant tidy, reduces contamination, and shortens cleanup time at the end of the production shift or when switching production from manufacture of one product to another. Employees who operate equipment that fills bottles or cans often have time to pick up debris, hose down spills, etc.

Beverage plants need standards for employee work habits and a training program to make sure employees meet the standards. Supervisors need to use frequent updates, training, and educational materials to make sure that good work habits and sanitation are priorities. Employees should be told how, when, and where to clean so that soil and debris are removed before they can provide food for pests and microorganisms. Leaky equipment should be fixed immediately. If rodents, birds, insects, or molds are found, employees should get rid of them, if possible, and report the problem to management. Employees should store products properly to discourage pests and allow easy cleaning. Employees should keep doors and windows closed.

The following are sanitation rules for beverage plants:

1. All employees visiting toilet facilities must wash their hands before returning to work.
2. Employees may not use spilled ingredients or products; they must throw them away.
3. Employees must put waste in containers with tight-fitting lids.
4. Each employee must keep his or her work area clean and tidy.
5. Employees may not use tobacco, except in designated areas, away from food-handling areas.
6. Employees may not spit anywhere in the plant.
7. Managers should inspect clothing, lunchrooms, and lockers regularly to make sure they are clean.
8. Employees must always wear head covers.

Cleaning Methods

There are five standard steps for cleaning a beverage plant:

1. Rinse to remove large debris and soil that is not stuck on, to wet the area and to help the cleaning compound work properly.
2. Apply a cleaning compound (usually as foam) to wet and penetrate the soil.
3. Rinse to remove the dispersed soil and the cleaning compound.
4. Sanitize to destroy microorganisms.
5. Rinse again if using a quaternary ammonium sanitizer (especially if the concentration is >200 ppm) before the cleaned area touches any beverage products.

NONALCOHOLIC BEVERAGE PLANT SANITATION

This book cannot discuss sanitation of all types of nonalcoholic beverage plants. Ultra-high-temperature (UHT) aseptic packaging is becoming more common and will probably be the "wave of the future," but this technology is too complex to cover here. A technical publication about aseptic technology can provide more information about this method.

Proper hygiene in a beverage-processing facility requires sanitary water, steam, air, carbon dioxide, and packaging. Many beverage processors use various types of filter to make sure that liquids and gases are free of microorganisms and other contaminants.

Beverages such as soft drinks, bottled water, beer, and distilled spirits need to be made from pure water, so manufacturers often need to treat the water. Processors use flocculation, filtration (through a sand bed), chlorination, sterile filtration, reverse osmosis, activated carbon, and deionization. The type of treatment depends on how the water will be used.

Beverage producers condition water by removing particles and controlling microbes. Particles are usually removed by flocculation and sand filtration. Microbes are controlled by chlorination.

Treatment with activated carbon removes excess chlorine, trihalomethanes, and other compounds used to disinfect the water. But activated carbon sheds fine particles of carbon that can provide sites for microbes to grow Also, carbon beds are difficult to disinfect. Therefore, the water is usually filtered again after leaving the carbon beds to reduce the load of microorganisms and particles.

Resin beds may be used to deionize water, but microbes can grow in the beds and beads of resin can get into the water. Again, the water is usually filtered after treatment in a resin bed. Sterile filtration does not use chemicals, is easy to use, and uses very little energy.

Steam is often used during production, but it can be a source of contamination. Steam is normally generated in carbon steel boilers, which rust easily. When a boiler operates continually, it develops a fine deposit of rust on the inside, which protects against further corrosion. When the boiler operates intermittently, fresh air gets inside and oxidizes the iron to iron oxides or rust. Rust forms continuously and flakes off into the steam.

Particles of rust from the boiler can damage equipment surfaces, block steam valves and filter pores, and stain equipment surfaces. Also, rust makes heat exchangers less efficient at heat transfer. To prevent this problem, processors often use culinary steam that is filtered using porous stainless-steel filters. The equipment has two parallel sets of filters so that one set can be cleaned while the other set is in use without interrupting the supply.

Many bottlers have installed cleaning-in-place (CIP) equipment to clean tanks, processing lines, and filters. Most bottlers that make several flavors of soft drink like CIP because it prevents the flavor "carry-over" from one drink to the next (especially from root beer).

One approach to cleaning beverage plants is known as TACT (time, action, concentration, and temperature). Using this method, it is possible to clean properly using different combinations of factors. For example, a 1% cleaning compound concentration at 43.5°C (110.3°F) may be the same as a 0.5% concentration at 60°C (140°F). However, it is important to remember that hotter water increases energy costs and condensation and cooks soil onto the surface being cleaned.

This section discusses cleaning in carbonated beverage plants, but other beverage plants use similar methods. Cleaning of floors, walls, and the bottling area are discussed under winery sanitation later in this chapter. Many cleaning methods in the beverage industry are similar to those used in dairy-processing plants (see Chap. 11).

Removing Tire Tracks

Tire tracks are difficult to remove. Alkaline cleaners work best. A mechanical scrubber helps remove the soil completely and more easily. Tire tracks are easier to remove and other soil will not be ground into the floor if the floor is cleaned every day.

Removing Soil from Conveyor Tracks

Soil on conveyor tracks is mostly spilled product, grease from bearings, filings from containers and tracks, and soap. A detergent track lubricant helps reduce contamination. Foam cleaning and a high-pressure rinse work well for cleaning conveyors.

Film Deposits

Films may form inside storage tanks, transfer lines, and filters. Thin films cause a dull surface, but more buildup looks bluish, and a thick film looks white. Residues from sugars are quite easy to remove, but aspartame and some gums make films that are difficult to remove. Processors may clean tanks manually, but circulation cleaning is more common. Chlorinated cleaning compounds work well for removing films.

Biofilms

Beverages and their ingredients have nutrients that allow microbes to grow and form biofilms. Biofilms can grow inside cooling towers, inside and outside warmers and pasteurizers, and inside coolers. As for films, chlorinated alkaline cleaning compounds work best to remove biofilms. A quaternary ammonium sanitizer helps stop biofilms from forming.

Hot Sanitizing

Sanitation in beverage plants is different from that in other food facilities. During the past few years, hot sanitizing has been used more and more. Hot sanitizing works well for surfaces that come in contact with beverages, such as batch tanks, flow mix units, fillers, and coolers. Hot sanitizing is not economical because it uses so much energy and does not remove bacteria well, but it does penetrate well. Heat is good at penetrating equipment and destroying microorganisms behind gaskets or in tiny crevices.

Hot sanitizing is not the same as sterilization. Hot sanitizing raises the surface temperature to 85°C (185°F) for 15 minutes; sterilization needs 116°C (241°F) for 20 minutes. Sanitizing reduces the number of microbes to a safe level, but a few of the most resistant microorganisms (yeasts and spores) stay alive. Chemical sanitizers can kill as many microbes as hot sanitizing but work much more quickly.

Special cleaning compounds can be added during hot sanitizing to loosen and remove soils and biofilms. These compounds are designed to remove soil, condition the water, and be rinsed off during hot sanitizing. Soil and biofilm removal are essential for good sanitation.

BREWERY SANITATION

The environment in a brewery restricts growth of pathogens and spoilage microbes. Bacteria found in breweries can cause several problems, such as changing the pH of the wort (the wort is the liquid that is fermented to make beer), making it more acid or alkaline, adding acetic acid, preventing complete fermentation, and making "ropes" in the wort. Bacteria may also cause off-odors and hazes in the finished beer.

Control of Microbial Infection

Undesirable bacteria, yeasts, and molds grow rapidly in freshly cooled wort if it is contaminated. Therefore, workers must clean and sanitize brewery equipment that processes the wort. High-speed equipment such as fillers, cappers, casers, and keggers perform better if they are kept clean.

The best way to prevent beverage products from spoiling is to develop and maintain a comprehensive cleaning and sanitizing program. Sanitation personnel, a reliable sanitation consulting group, or a dependable cleaning compound and sanitizer supply firm can develop the program. Chapters 5 to 7 give more information on this topic. CIP (Chap. 7) is used more and more often for cleaning beverage equipment.

Fermentation facilities such as breweries need sterile air to make starter cultures and maintain sterile conditions inside storage tanks.

Cleaning Compounds

It is important to use the right cleaning compound for the type of soil. When spray cleaning, the cleaning compound should not make foam, because foam stops the solution from touching every part of the surface. The right cleaning compound will not allow beerstone (a deposit that can form on brewery equipment) to form. The cleaner should not attack metal and must be easily rinsed away so that the taste of the cleaner does not get into the beer. (See Chap. 5 for more information about cleaning compounds.)

Sanitizers

Chlorine or iodine are good sanitizers for fermenters, cold wort lines, and coolers. These sanitizers are added with the final rinse. Water can contain microorganisms; therefore, the final rinse should contain a sanitizer so that the final rinse does not leave bacteria or yeasts on the sterile surfaces of the equipment. (See Chap. 6 for more information about sanitizers.)

Heat Pasteurization

Producers of packaged beer use heat pasteurization to control microbes. Heat pasteurization is convenient, but it uses a lot of energy, which makes it expensive. Beverage producers have tried different methods because heat pasteurization:

*   Has high energy costs
*   Alters the flavor of beer
*   Requires equipment that occupies space
*   Makes working areas hot and humid

The alternative is cold pasteurization, using either:

*   Chemicals, such as propyl gallate
*   Millipore filtration and aseptic packaging or millipore filtration
    and other chemical treatments

New technology and safety information affect whether various chemical treatments are officially approved.

Aseptic Filling

Aseptic filling is different from pasteurization. Aseptic filling uses ultrafiltration to remove spoilage organisms from beer before packaging. Ultrafiltration occurs before packaging, so spoilage microorganisms can still get into the beer before it is sealed. The following suggestions help protect the product during aseptic filling:

*   Filling rooms must be closed and have positive pressure of filtered air.

- Workers' clothes should always be clean.
- All involved should wash their hands with a sanitizing hand soap before entering the room.
- The conveyor belt should have a lubricant system that reduces microorganisms.
- The inside of the filler should be cleaned and sanitized daily, using recirculating CIP equipment.
- The outside of the filler, conveyors, connected equipment, floors, and walls must be foamed or gelled and sanitized every day. The detergent or sanitizer that is left on the surface after cleaning should prevent microbes from recontaminating the sanitized surfaces.
- Surfaces and air in the filling area should be monitored regularly to check for bacteria, yeasts, and molds. A HACCP (Hazard Analysis Critical Control Point) program can include chemical and microbial monitoring to guarantee safe food production. The results should be checked to make sure levels of contaminants do not go too high and that levels do not slowly increase.

Bottle Cleaning

Centralized high-pressure, low-volume cleaning equipment has improved bottle cleaning (see Chap. 7). This method can remove soil from very hard-to-reach equipment such as conveyors, bottle fillers, cappers, and casers.

Sanitation in Storage Areas

Edible dry products—such as grain, sugar and other dry products—need proper storage. Workers should clean screw conveyors (used to carry dry ingredients) regularly. Staff must be careful to clean the dead ends of conveyors where residues collect. They should clean the ends and joins between conveyors at least once a week. The free-flowing section of a conveyor should have hinged covers for easy cleaning and inspection. After conveyors are cleaned, workers should fumigate them with a fumigant that does not leave a residue. They should sweep and vacuum empty bins before fumigating them. Sanitarians should check material that is cleaned out of bins to make sure it is not infested. (Chap. 9 discusses pest control in dry storage areas.)

Brewing Area Sanitation

Spray cleaning is faster and works better than manual cleaning and can reduce the amount of time production has to stop for cleaning. Cold water can be used, but a water temperature up to 45°C (113°F) can help the cleaning compound react with the soil. In glass-lined tanks, the water should not be hotter than 28.5°C (83°F) because sudden temperature changes can break the glass. Temperatures hotter than 45°C (113°F) are not recommended because hot water causes condensation and products need more refrigeration. Management should install high-temperature cutoff switches to control water temperature. Caustic-soda cleaning compounds attack soldered equipment and should not be used. Scale can be removed from aluminum vessels using 10% nitric acid, applied as a paste mixed with kaolin.

Hoses and fittings are expensive to buy and maintain. In the long run, stainless-steel lines are more cost-effective because they are so durable, even though stainless steel is quite expensive initially. Industrial spray nozzles can be put in places where they can clean vapor stacks on kettles, strainer troughs in hop strainers, and conveyor belts. Workers should clean the brewing area at least once a week and should remove debris and other obvious dirt every day.

Beerstone is mainly organic matter in a matrix of calcium oxalate and is one of the most difficult beverage soils to remove. A strong chelating agent, an alkaline cleaning compound, and heavy scrubbing work best.

Bottle Washing

Workers must carefully examine all empty bottles when they are returned by the retailer and the bottler. They should also inspect new bottles to make sure there are no obvious signs of contamination. All new and used bottles should be washed immediately before filling. A mechanical washer that sprays plenty of caustic solution on the inside and outside of the bottle works best. The temperature of the water used for spraying and rinsing should be 60 to 70°C (140–158°F). The final rinse can contain up to 0.5 part per million (ppm) of chlorine without affecting the flavor of the beer. But chlorination is necessary only if the rinse water is not pure.

Beer Pasteurization

Most brewers pasteurize their beer to make sure the condition, flavor, and smoothness do not change during distribution. Some brewers use sterile filtration instead of sterilization. If used, staff should replace filters every other week so that microbes cannot grow in the filters.

Beer is often pasteurized when it is bottled or canned because it is unlikely that the beer will be contaminated again after it is sealed. Overheating during pasteurization can spoil the flavor and make the beer hazy. So it is important to use the shortest amount of time and the lowest temperature possible while still destroying enough microbes. Most breweries now pasteurize using a conveyor system with a pasteurization cycle that takes about 45 minutes. During pasteurization, the beer is heated slowly from 1 to 2°C (34–36°F) to 61 to 63°C (142–145°F) and then cooled to room temperature. The speed of the conveyor belt controls how long the beer is heated.

Haze in beer may be nonbiological or biological. Nonbiological haze forms when products that are not very soluble settle out of the beer. This process is quicker when oxygen (air) is present. Biological haze forms when bacteria or yeasts grow in the beer. If the beer stays long enough in the cold conditioning tank and goes through fine filters it is unlikely to have nonbiological haze. Keeping air out of the beer container and choosing the right materials for containers also make nonbiological haze unlikely. Other hazes can be caused by metal, especially tin. If beer has a haze due to bacterial or yeast growth, it was probably not filtered properly or was infected after filtration. A bacteria or yeast haze can be caused by poor sanitation in the plant or unclean storage containers or filters.

Cleaning Air-Conditioning Units

A good procedure for cleaning air-conditioning units follows:

1. Clean air-conditioning units every 6 months. Insert a ball spray through a special opening above the coils on top of the air-conditioning units.
2. Run water through the ball spray for 10 minutes to flush the unit.
3. Run a hypochlorite solution (200 ppm) at 40°C (104°F) through the ball spray for 5 minutes.
4. Let the unit soak for 5 minutes.
5. Rinse the unit with warm water for 10 minutes.
6. Check the unit and clean the pan bottom.

Conserving Water in Breweries

A wash-rinse cycle, sometimes called the *slop cycle,* can help breweries use less water during cleaning. This cycle is similar to the one used in most home dishwashers. A cleaning solution is pumped through a spray device for 20 seconds, left to soak for 1 minute for the cleaning chemical to work, and rinsed with water. The system is practical and economical because the cleaning solution is reused. The solution can be reused for longer if the tank that holds the cleaning solution has a top overflow to skim off floating soil and a drain valve at the bottom so that heavy soil can be drained out. The final rinse water can also be reused to prerinse another tank. This technique can reduce water and sewage treatment costs in areas that meter water use and sewage production.

WINERY SANITATION

Winemakers must remove any contaminants that spoil the taste, appearance, and keeping quality of wine. Several things can contaminate wine—for example, the reddish tartrate deposits that build up inside tanks during fermentation. It is important to clean processing equipment carefully to reduce growth of microbes everywhere in the winery. Wineries add sulfur dioxide (SO_2) at the end of the winemaking process to stop it from fermenting, but sanitary wineries only need to add very small amounts.

Sanitation becomes more and more important during the winemaking process and is most important at bottling time, but even the vineyard tools and harvesting equipment need to be washed to remove dirt, soil, and leaves. Destemmers, crushers, grape-processing areas, and bulk storage areas need brushing with detergent and water. Harvester heads, pipes, hoses, pumps, faucets, spigots, and anything else that touches the juice or wine should be cleaned according to the five cleaning steps discussed earlier in this chapter. The same steps are used in the bottling line, but with extra control and checking to control the number of microbes in the wine.

The water used in a winery must meet chemical and microbial standards. Water can carry molds, yeasts, and acetic or lactic acid bacteria, so it is important to use pure water.

The design and layout of the winery should support good hygiene. Floors must be easy to clean with a nonslip, sloped surface. Walls and ceilings should be sealed and easy to clean. Winemakers should be careful to place equipment so that it does not create corners and crevices that are difficult to clean and so that floors are easy to clean. The equipment should have the same sanitary construction features as in other food-processing plants.

Cleaning Floors and Walls

Operation of a winery may be seasonal, but it needs to be sanitized year-round. Wet and dry cleaning methods are usually used together, depending on the type of soil. A heavy-duty, wet-dry vacuum cleaner can be used for floors, which should be cleaned at least once a week. Floors should be made of concrete and sloped, and should contain trench drains. Spilled wine, especially spoiled wine, should be washed away immediately. The area of the spill should be mechanically scrubbed, washed with lime or a strong hypochlorite solution, and rinsed with water. Floors should be periodically washed and disinfected with a dilute hypochlorite solution. Dry cleaning helps to keep the humidity of the winery low, which reduces mold growth. Tank tops, overhead platforms, and ramps can be vacuumed, cleaned, or washed, making sure that no water gets into the wine. The walls should be washed with a warm alkaline solution (such as a strong mixture of soda ash and caustic soda), rinsed with water, and sprayed with a hypochlorite solution containing 500 mg/L (0.67 oz/gal) of available chlorine.

Equipment Cleaning

Improperly cleaned equipment is likely to contaminate the wine. Crushers, pumps, lines, presses, filters, hoses, pipes, and tank cars are all difficult to clean completely.
Equipment should be:

- Taken apart as much as possible
- Thoroughly washed with water and a phosphate or carbonate cleaner for non-metallic surfaces, and caustic soda or equivalent for cleaning metal equipment
- Sanitized with hypochlorite or an iodophor if the material being cleaned is adversely affected by chlorine

Enzymes are useful as cleaning agents because they can break down proteins, fats, and pectins (found on fruit). Enzymes are used in enology (winemaking) because they work best when the pH is near neutral (not acid or alkaline). Whenever possible, circulating the cleaning solutions is recommended. After cleaning and rinsing, hoses should be placed in sloping racks, rather than on the floor, to help them drain and dry. Equipment that has touched spoiled or contaminated wine must be thoroughly cleaned and sanitized.

During the harvest season, conveyors, crushers, and must (juice) lines should be kept clean. They should not stand with must in them for more than 2 hours. They should be washed, drained, and thoroughly flushed with water every 2 days.

Bottling-Area Cleaning

Good cleaning of the bottling area reduces contamination by bacteria and metals. Public-health agencies usually watch this area closely. The room should have good lighting and ventilation, glazed tile walls, and epoxy finished floors. Equipment should be far enough apart to allow cleaning in between, and the equipment should be easy to take apart. All pumps, pipes, and pasteurizers should be made of stainless steel so that debris, mold spores, and yeasts from freshly cut corks can be removed. Corks should be cleaned and sterilized before use by soaking for 2 hours in 1% sulfur dioxide and a little glycerol; and then rinsed with water. Corks can be sterilized using gamma-radiation to stop mold growth from causing off-odors.

Pomace Disposal

It is essential to get rid of the pomace (dry grape skins after the juice is squeezed out) as soon as possible after the grapes are pressed. Pomace must not stand in or near the fermentation room because it has a high acetic acid content, and fruit flies carry acetic acid bacteria from the pomace pile to clean fermenting vats. Pomace should be processed or scattered as a thin layer on fields. This thin layer dries quickly, so it does not become a serious breeding ground for fruit flies.

Cleaning Used Barrels

Alkaline solutions work best for removing tannins from new barrels. Barrels may be soaked in 1% sodium carbonate. If they need more cleaning, two atmospheres pressure of heated steam and several rinsings can be used.

In the past, empty barrels were washed with water and sprayed with a hot (50°C, 122°F) 20% solution of a mixture of 90% soda ash and 10% caustic soda. Then they were washed with hot (50°C, 122°F) water, sprayed with a chlorine sanitizer solution (containing 400 ppm of available chlorine), rinsed, drained, and dried using a dry-wet vacuum.

Any mold should be scraped off; because washing does not remove it. Washing with a quaternary ammonium compound and paints containing copper-8-quinolate help control mold growth. Burning a sulfur wick in the tank or the equivalent amount of sulfur dioxide from a cylinder of gas also works well. Before use, the tanks should be rewashed, and the barrels should be inspected and smelled before being filled. A warm 5% soda-ash solution neutralizes vinegar in used barrels. This treatment should be followed by wet steaming and washing. If the soda-ash concentration is too high and the barrel is exposed for too long, it can damage the wood. The outside of wooden containers should be washed with a dilute solution once a year. Propylene glycol (400 ppm or less) can be used to control mold growth on tanks.

Removing tartrate deposits. Tartrate deposits make the inside surface of barrels very rough. Scraping is labor intensive and may injure the wood. A circular spray head inside the tank can help remove tartrates. Soaking with 1 kg of soda ash and caustic soda in 100 L of water also helps remove tartrates.

Storing Empty Containers

Concrete tanks should be left open and kept dry when not in use. They should be inspected and cleaned before they are used again. Open wooden fermenters are sometimes painted with a lime paste when they are not being used, but this is difficult to remove. It is better to clean the fermenters with an alkaline solution and then a chlorine solution. They can then be filled with water and about 1.6 kg of unslaked lime per 1,000 L (260 gal) of water. Stored empty barrels can be sulfured using a sulfur wick or SO_2.

Other Cleaning and Sanitation

Fillers, bottling lines, and other packaging equipment can be cleaned with CIP systems. A chlorinated alkaline cleaning compound can clean, sanitize, and deodorize at the same time if the soil is light. But the chlorine sanitizer cannot work if the equipment still contains organic matter.

Heat is the safest sterilization method. It can be used for everyday wines, but it is not good for high-quality premium wines because heat can change the taste. Sterilization in the bottle works best for sparkling wine.

Sterile Filtration

Filtration can sterilize wine via sterile filter pads or membranes. Diatomaceous earth filtration can reduce the yeast population but does not remove bacteria. This filtration can be followed by membrane filtration.

Reinfection

Most problems are caused by improper sterile filtration. Sterile filtration is ruined if the whole bottling line is not sterile. The best way to sterilize the whole system is to use a steam generator hooked to the filter and bottling line. A slow flow of low-pressure steam should run through the whole system for 30 minutes. The equipment is cooled using cold water before the wine is filtered and bottled.

Steam is not always available, and it can damage some equipment, such as plastic filter plates. Citric acid (300 g/hl [0.4 oz/gal] and 10 g/hl [0.013 oz/gal] SO_2 or

20 g/hl [0.027 oz/gal] metabisulfite) at a temperature of 60°C (140°F) is an alternative way to sterilize the system.

Some parts of the bottling line, such as the corker, are more difficult to sterilize. The corker jaws or diaphragm should be sterilized with alcohol. Membrane filters may be sterilized with water at 90°C (194°F).

Corks

Modern cork suppliers provide sterile corks. If winemakers are uncertain whether they are sterile, they should dip them in a 10 g/hl of SO_2 [0.013 oz/gal] solution before use.

Bottles

Bottles normally come in cases and can be contaminated by dust and cardboard. Bottles must be rinsed and sterilized using a solution of 500 ppm SO_2. Excess SO_2 is washed off the bottle using sterile water. An SO_2 dispenser can be placed on the main water supply.

Control

Sanitation standards must be checked during bottling. Winemakers can get special kits to check sanitation by counting the number of live yeasts or spoilage bacteria left in the wine after filling.

Pest Control

Fruit flies are attracted to fermenting musts. Many flies come to the winery from the vineyard. To control flies, crush grapes promptly after picking, remove all dropped and damaged fruit from the winery, get rid of all organic waste, use insecticides around the vineyard, wash all containers and trucks after handling grapes, and use insecticides on dumps. Flies are most active between 23.5 to 27.0°C (74–81°F), in dim light, and with little wind. Fans to blow air out of winery entrances, mesh screens, and air curtains help keep flies away.

Insecticides can kill fruit flies, but they only work on the area sprayed, and nearby areas may still have large fruit fly populations. If insecticides are used, they must be used at levels below the U.S. Food and Drug Administration tolerances. (Chap. 9 provides additional information related to fly, rodent, and bird control.)

DISTILLERY SANITATION

Sanitation Notes

As in breweries and wineries, the microorganisms that cause most problems in food-processing plants rarely cause problems in distilleries because of the ingredients, processing techniques, and high alcohol concentration. However, distilled beverages can be contaminated. It is important to control the raw materials, because a contaminated finished product cannot be detoxified. When conditions in distilleries are not sanitary, the yield and the quality are poor.

Reducing Physical Contamination

Corn and other grains are inspected when they arrive at the plant. The biggest concern is insects, because a contaminated grain shipment can infect the storage silos and the entire plant. The most common insect pests in grain are flour beetles and weevils. It is also important to check grain for off-odors, because these can carry over into the final product. Grain storage silos are emptied 2 to 4 times a year, sprayed with high-pressure hoses, and allowed to air-dry. The area around the silos is kept clean by washing the area with water and spraying with insecticide.

Grain enters the plant on conveyors and should be sifted on shaking screens to remove corn cobs, debris, or insects. The mill room should be washed down with water to reduce grain dust, and it should be heated to 55°C (131°F) for 1/2 hour every 2 to 6 months to kill insects.

Reducing Microbial Contamination

Bacteria and wild yeasts can contaminate whiskey during fermentation. Most of the contamination comes from the malted barley. The malt is added at 60 to 63°C (140–145°F), so many of the microbes survive and grow during fermentation. Bacteria in the corn and other grains that are added before the cooking process or at temperatures greater than 88°C (190°F) are not a problem, because they are killed by the high temperatures.

Many microbes cannot grow during the fermentation process. At the beginning, sugar concentrations may be above 16%, and the pH is between 5.0 and 5.4 (acidic) for a sour mash whiskey. At the end the pH is between 4.0 and 4.5 (more acidic). Final alcohol concentration is about 9%, and the mixture contains plenty of carbon dioxide but very little oxygen. These conditions restrict the types of organisms that can grow.

To make sure the beverage is not contaminated during fermentation, dust is kept down by washing all plant surfaces (walls, floors, etc.) with water, and incoming shipments of malt are tested to check the level of bacteria. (Most distillers limit total bacterial counts to between 200,000 and 1,000,000/g.)

Equipment Cleaning

Large fermentation vessels (114,000–171,000 L, 30,000–45,000 gal) are cylindrical tanks with cooling coils on the inside. They should be cleaned by filling with hot water, detergent, and steam using a CIP system. After 30 minutes, the tank should be emptied, rinsed with water, and steamed for 2 to 3 hours to sterilize before new mash (crushed malt or grain) is pumped into the vessel.

The cooling coils should keep fermentation temperatures below 32°C (90°F). At hotter temperatures, the yeasts that cause fermentation die, and off-flavors develop. Cooling coils may collect "beerstone," which is hard and rocklike and contains calcium carbonates, phosphates, and sometimes sulfates. When beerstone builds up on the coils, they become less efficient at cooling. The beerstone should be removed every 6 months by filling the vessels with a 1% caustic solution (sodium hydroxide) and soaking for 3 days.

Residual grain builds up in the cookers, where the mash is prepared, and in the beer still, where the finished beer is pumped. The cookers, the beer still, and all connecting lines should be washed weekly with a 1% caustic solution. Some distillers use a 1% solution of acetic acid instead. The caustic solution can be prepared in a large tank and pumped to the beer still through the connecting lines and into each of the cookers. These areas should be rinsed with water to wash away all caustic residues.

The lines and stainless-steel tanks that carry distilled alcohol to receiving tanks and then into maturation barrels for maturation should be rinsed periodically. Because the product is crystal clear alcohol at 140 proof, sanitation does not need to be stricter.

Distillery products must be made using good-quality water. Water used to blend distilled spirits usually comes from a chlorinated and carbon-treated well or city water supply and is filtered. Chlorination makes the water microbiologically safe. Filtering makes sure the water is completely clear (no cloudiness).

SUMMARY

- Most soils in beverage plants contain a lot of sugar, dissolve in water, and are easy to remove.
- Microorganisms cannot be removed from the environment, so good sanitation is essential to control them.
- Strict control of raw materials is essential, because a beverage producer cannot detoxify a contaminated finished product.
- The bacteria that cause most problems in breweries are non-spore-formers. The best way to make sure beverage products do not spoil is to control contamination through a complete cleaning and sanitizing program.
- Spray cleaning works best with a properly blended, low-foaming cleaning compound, designed for the type of soil in the plant.
- Sanitizers—such as chlorine, iodine, or an acid-anionic surfactant—are recommended with the final rinse in fermenters, cold wort lines, and coolers.
- Sanitation becomes more and more important during the winemaking process and is most important at bottling time. Wet and dry cleaning are used together.
- Winemaking equipment should be taken apart as much as possible, thoroughly washed with water and a phosphate or carbonate cleaner for nonmetallic surfaces and caustic soda or equivalent for cleaning metal equipment, and sanitized with hypochlorite or an iodophor.
- Ways to remove tartrates include installing a circular spray head inside a tank or soaking with soda ash and caustic soda.

- Fillers, bottling lines, and other packaging equipment can be cleaned with a CIP system.
- Prompt processing of grapes after picking reduces fly infestation.

BIBLIOGRAPHY

Amerine, M. A., and Joslyn, M. A. 1970. *Table Wines: The Technology of Their Production.* 2d ed. University of California Press, Berkeley.

Amerine, M. A., Berg, H. W., Kunkee, R. E., Ough, C. S., Singleton, V. L., and Webb, A. D. 1980. *The Technology of Winemaking,* 4th ed. AVI Publishing Company, Westport, Conn.

Arnett, A. T. 1992. Distillery sanitation. Unpublished information.

Connolly, B. J. 1971. Stainless steel for wine storage and treatment installations. *Rev. Vin. Intern.* 92(180): 49.

Hough, J. S., Briggs, D. E., and Stevens, R. 1971. *Malting and Brewing Science.* Chapman & Hall, London.

Jefferey, E. J. 1956. *Brewing Theory and Practice,* 3d ed. Nichols Kaye, London.

Marriott, N. G. 1994. *Principles of Food Sanitation,* 3d ed. Chapman & Hall, New York.

O'Sullivan, T. 1992. High quality utilities in the food and beverage industry. *Dairy Food Sanit.* 12:216.

Remus, C. A. 1991. Just what is being sanitized? *Beverage World* (Mar.):80.

Remus, C. A. 1991b. When a high level of sanitation counts. *Beverage World* (Mar.):63.

Remus, C. A. 1989. Arrhenius' legacy. *Beverage World* (Apr. 1991):76.

Stanton, J. H. 1971. Sanitation techniques for the brewhouse, cellar and bottleshop. *Tech. Q. Master Brew. Assoc. Am.* 8:148.

STUDY QUESTIONS

1. Why are soils in beverage plants easy to remove?
2. Why is it important to keep beverages sanitary throughout processing?
3. What are the five standard steps for cleaning a beverage plant?
4. How do beverage producers purify water?
5. List one advantage and one disadvantage of sanitizing using heat.
6. Which part of beverage production needs the strictest sanitation?
7. What characteristics protect beverages from microbes?

TO FIND OUT MORE ABOUT BEVERAGE PLANT SANITATION

1. Contact a local winery, brewery, or distillery, and arrange a tour. Ask about their sanitation procedures.
2. Contact a soft-drink manufacturer (many have toll-free telephone numbers listed on the can or bottle). Ask where they obtain the water used in their soft drinks and how the water is treated.

GLASS OR PLASTIC BOTTLES FOR MINERAL WATER?

By law, natural mineral water should contain only the types of bacteria it contains at the spring. It is very difficult to keep other bacteria out of the water, especially when the water sits in large tanks and reservoirs before it is bottled. Bottling also affects the number and type of bacteria in the water, because the open system (open to the air) becomes a closed system (without air). The type of bottle (glass or plastic) also affects the bacteria. In glass bottles, the number of bacteria in the mineral water goes down after bottling. This is probably because residues of cleaning detergents in the bottles inhibit growth of bacteria. Organic substances that are dissolved in the plastic bottles have very little effect on bacteria in the water. Beverage producers bottling natural spring water need to be very careful to make sure that water is not contaminated with pathogenic bacteria and that these cannot grow in the water after it is bottled.

Source: Bischofberger, T., Cha, S. K., Schmitt, R., König, B., and Schmidt-Lorenz, W. 1990. The bacterial flora of non-carbonated, natural mineral water from the springs to reservoir and glass and plastic bottles. *Int. J. Food Microbiol.* 11: 51–72.

Low-Moisture-Food Sanitation

ABOUT THIS CHAPTER

In this chapter you will learn:

1. Why good sanitation is important in low-moisture food-processing facilities, even though microbes do not usually grow in these foods
2. About the pests, insects, and other contaminants that affect low-moisture foods
3. How the design of a low-moisture-food plant and the type and layout of equipment affect sanitation
4. Why and how employees should inspect low-moisture foods when they arrive at the plant and while they are in storage
5. How to store low-moisture foods
6. How to clean low-moisture food plants

INTRODUCTION

Low-moisture foods include:

- Bakery goods—bread, rolls, muffins, cookies, cakes, pies, pastries, doughnuts, etc.
- Nuts and seeds
- Candy
- Cereal—ready-to-eat cold cereal and cereal to be eaten hot (e.g., oatmeal)
- Grains—flour, cornmeal, rice, dry pasta, baking mixes, etc.
- Snack foods—chips, crackers, pretzels, popcorn, etc.

These foods are called low-moisture foods because they contain very little water, usually less than 25%. This means that most bacteria do not grow in these foods. Even though these foods are less likely to spoil than dairy products, meat, poultry, fish, fresh fruits and vegetables, and beverages, they can still be contaminated and need hygienic processing and storage.

Just like other food plants, low-moisture-food manufacturing plants need a good sanitation program. The plant must meet the requirements of the U.S. Food and Drug Administration (FDA), the state, and local authorities. Rigid sanitation is needed to make sure that consumers buy safe and wholesome foods.

Bakeries, bottlers, and food warehouses are usually thought of as low risk for food contamination. But the Office of the Inspector General of the Department of Health and Human Services has said that these facilities are becoming more risky because of poor inspection. FDA regulates firms that ship products between states. State and local authorities inspect their facilities, but inspections are rare and look mainly for problems with birds, rodents, and insects. Firms with unsanitary conditions are a public-health risk. Low-moisture-food operations need a good sanitation program to make sure that customers get wholesome foods.

Tidiness and cleanliness help food operations to be efficient, help promote a good image for the company, and affect whether an operation makes a profit or even stays in business. Poor sanitation leads to dissatisfied customers, decreased sales, and a bad reputation for the firm.

FACILITY DESIGN AND CONSTRUCTION

Choosing a Site

A good location for a low-moisture food-processing plant has the following features:

- Level ground or only a slight slope with good drainage
- No springs or areas where water collects
- Available municipal services (sewage, police, and fire)
- Access to major railways and highways in all weather and seasons
- Large enough to double the size of the production facilities and truck parking area
- Not close to areas scheduled for redevelopment or highway construction
- Good supplies of inexpensive and good-quality water, natural gas, electricity, fuel, and other utilities
- Not close to incinerators, sewage treatment plants, and other sources of bad smells, pests, or bacteria
- In an area where gases produced during heating and baking are allowed to be released into the air
- Away from areas that flood, have earthquakes, or have other natural disasters
- Near a good labor market for current and future staff needs

Design of the Outside

The outside of the building should look clean and neat and give a good impression to employees and visitors. The walls should be smooth and waterproof and should not have ledges or overhangs where birds can live. The builder should seal the walls against rodents and insects. Driveways should be paved and have no weeds growing on them. Trash and water should not collect on driveways. Staff should sweep the outside area regularly so that dust does not blow into storage areas.

Design of the Inside

Walls and framing. Exposed pipes and structures may be fine in areas where food is not processed or stored, as long as they are kept clean and dust-free. Reinforced concrete construction is best for areas where food is processed or stored. The area should have as few inside columns as possible. Staff doors should have hydraulic or spring hinges so they close by themselves, and the doors should have screens. Gaps at the bottom of doors should be less than 0.6 cm (0.25 in), and screens should be at least 20 mesh.

Walls must not have cracks and crevices and must not soak up water or other liquids, so that they can be cleaned easily and thoroughly. Wall finishes should be appropriate for each area. Glazed tile is a good surface for walls in the processing area. Another alternative is fiberglass-reinforced plastic panels, painted with epoxy or coated with other materials that meet company and regulatory standards. Paint is cheap but is not good for food areas, because it cracks, flakes, and chips after a while and needs to be redone often.

In bakery facilities, insulation can be a home for insects. The builder should place insulation on the outside of the building so that pests cannot live in it inside the building.

Ceilings. Suspended ceilings are fine in areas where food is not processed or stored, so long as it is possible to inspect the space above the ceiling and keep pests, dust, and other debris out. The builder must seal the ceiling panels into the grid, but workers must be able to remove them easily for cleaning.

Suspended ceilings are not suitable for food production or handling areas. Suspended ceilings can mold if they get wet and provide a place for pests to live. In areas of bakeries that use flour, dust collects above the ceiling very quickly and can cause fires or explosions, as well as provide food for pests and microbes.

Whenever possible, building designs should not use overhead structures such as joists and beams. Precast-concrete roof panels form a clean, clear ceiling. Precast panels can be made with a smooth inside surface, coated to resist dust, and easily cleaned. Supports for overhead equipment; gas pipes; water, steam, and air lines; and electrical wires should not pass over areas where food is open, clutter the ceiling, and drip dust or water onto people, equipment, or food. A mechanical mezzanine floor can be used for overhead utility equipment. It leaves the ceiling easy to clean and free from horizontal pipes and ducts (see Fig. 16.1).

Floors. Floors in areas that are washed with water must not soak up water, must not have cracks and crevices, or react with chemicals and acids. Yeast foods (e.g., yeast mixtures used to make bread) are very corrosive. Builders must seal floor joints and

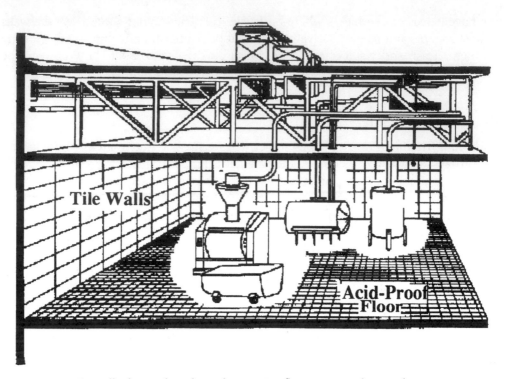

FIGURE 16.1. Specially designed mechanical mezzanine floor separates ducts and support structures from the bakery mixing room. This cuts down on overhead cleaning, makes it easier to get to equipment for maintenance, and protects food products.

cover and seal wall junctions. Whenever possible, builders should use expansive concrete to cut down on the number of joints. In 1986, the U.S. Department of Agriculture recommended that floors should have a slope towards drains with about a 2% grade. This allows proper drainage after wet cleaning. Processing equipment should connect to drain lines, and equipment should have drip pans to cut down on floor spills.

Storage areas should have a white stripe 0.5 meters (18 in) wide painted around the edge. Nothing should be stored in this area. Food must be stored away from non-food items. Some products must be kept separate so that they do not contaminate each other, for example, foods and chemicals.

Different types of operation need different floors. Reinforced concrete, coated or hardened to keep away dust, may be fine for packaging and oven areas of bakeries. But areas used for liquid fermentation and handling dough should have a surface that will not be damaged when it comes in contact with hot water, steam, acids, sugar, and other ingredients or sanitizing chemicals during food processing and cleaning.

Chemical-resistant floors are best for areas that are cleaned using water. Epoxies, polyester, tile, or brick are all good. The builder should bond toppings right onto the base, such as concrete. Toppings give a watertight barrier that protects the concrete.

"Dairy" tile or pavers are good for areas with heavy traffic or for surfaces that touch food products or cleaning solutions. Builders should install dairy tile with acid-resistant bonds. It is very durable and sanitary. It is easy to clean and can be manufactured with a nonslip finish. It is expensive, but it can save money in the long run.

Floors in special areas, such as coolers and freezers, need to be made of the best materials for the area and need good insulation and ventilation. If the freezer floor does not have insulation, it can freeze the ground under it, which can crack or buckle the freezer floor and jam the doors.

Ventilation and Dust Control

It is very important to control dust. The manufacturing process heats the food above pasteurization temperatures and usually kills live microbes. But spores can survive inside baked goods, especially those that are soft and moist. Also, dust from raw materials can contaminate finished food products.

Processors must design facilities so that finished foods are not contaminated. Processors must consider sanitation procedures, the arrangement of the equipment, ventilation, and dust control. They must also control the temperature and humidity so that bacteria cannot grow quickly.

Planning Equipment

The type of equipment and how it is arranged affect productivity. A mechanical mezzanine floor should separate heating and cooling equipment from processing areas. The plant should have high-efficiency motors and electrical equipment (Fig. 16.1), the best automatic technology and controls that the processor can afford, and a flexible design to respond to changes in processing and consumer demands. All equipment should meet current regulations.

Sanitation

Like other food-processing facilities, the building is there to protect what is inside. The building must protect the process, equipment, and food products. Sanitation, productivity, product flow, and choice of equipment and site are all very important.

An operation such as breadmaking involves fermentation. Fermentation uses yeast, and the bakery must be kept sanitary for the right organisms to grow. Poor sanitation allows natural microorganisms to ferment the dough instead of the proper yeast. Once they are growing in the facility, natural microorganisms are very difficult to remove and difficult to control.

Current Good Manufacturing Practices (CGMPs) related to design and construction should:

1. Allow enough space for equipment and storage
2. Separate operations that might contaminate food

3. Provide enough light
4. Provide enough ventilation
5. Protect against pests

It is best to keep the inside of the plant simple and uncluttered. Designers should design the plant layout with sanitation in mind. The design team should include technical staff, engineers, and production staff from the plant, as well as a contracted design and engineering firm.

Productivity

The well-designed, efficient plant makes a productive operation. A good design should have:

- Flexibility to allow for future changes
- Enough space to process a larger volume to allow for growth
- Dry and refrigerated storage areas in convenient places
- A central plant management office and laboratory so that managers can supervise all processes closely
- Ingredient storage areas near where they are mixed and used
- Secondary equipment, such as boilers and refrigeration machines, in places that minimize pipes and utilities
- Open space around and under equipment for cleaning and maintenance
- Manufacturing equipment arranged for easy cleaning in place (CIP)
- Proven or tested equipment and processes
- State-of-the-art controls and automation; facilities should update systems as technology advances and becomes less expensive
- Conveniently located restrooms, lockers, and break areas away from work areas
- Plenty of handwashing stations in work areas
- A computer simulation of the proposed layout, if possible
- Extensive design input from the plant's engineering and operation staff to make sure it complies with the company's standards and will not cause problems in day-to-day operations
- Compliance with federal, state, and local regulations

RECEIVING AND STORING RAW MATERIALS

Checking Raw Materials

Staff cannot check all of the raw materials that come into the plant. They need a system for collecting samples to decide if they should accept or reject deliveries. They need to collect enough samples (a statistically valid size and number) to be confident that the raw materials are acceptable.

Inspecting Trucks

Trucks used to transport low-moisture raw materials before, during, and after un-loading should be inspected. Inspectors should look at the overall condition of the vehicle, areas where food and dust can collect, and potential homes for insects. It is important to check for insects in the area around doors or hatches. Inspectors should look for crawling or flying insects, animal tracks, nesting materials, smells, and feces. Pellets and smells are signs of rodents; feathers or droppings are signs of birds.

Evaluating Raw Materials

Inspectors should check the moisture content of raw materials. They can analyze for percentage moisture, but can also check for signs of high moisture content. A sour or musty smell can be caused by mold, which can only grow on products such as cereal grains when they are moist. If inspectors think the material may be moldy, they should do more tests. Cereal grains with more than 15% moisture should not be stored for long because they are likely to mold and attract insects.

Inspectors should also check that raw materials do not smell of pesticides that may have been used to get rid of insects. The inspector may need to test the food to find out if the pesticide levels are too high.

Inspectors should check samples of raw materials to see what percentage, if any, has been destroyed by insects or rodents. They should also check the amount of dust, other contaminants, insect webs, molds, smells, live and dead insects, and rodent droppings or hair. Immature insects may be inside grains; X-ray machines or cracking-flotation methods can test for these.

Inspecting goods when they arrive reduces pest damage and helps prevent raw materials from contaminating the end product. Pests can "hitchhike" into buildings and settle on ingredients, packages, pallets, and machinery, so inspectors need to check all of these. A food processor has the right to reject any materials that come to the plant or hold them for more tests. Trained and experienced personnel should decide to accept or reject shipments.

Storing Food

Many low-moisture-food-processing plants store material such as grain before processing it. The United States stores more grain than any other commodity. But grain contains mold spores and insect eggs that grow and damage the grain under some conditions. If grain kernels are physically damaged, insects and molds can get into them and spoil them more easily. Insects can also cause biological damage to kernels, which allows molds to get inside.

If a processor is going to store grain for more than 1 month, the grain needs special treatment:

• Inspectors should make sure that it is not already infested or infected.
• Moisture content must be less than 13.5%.

- Processors may clean grain before storage, using aspiration or other methods to remove insects, weed seeds, and other foreign materials.

- Processors may use chemicals to protect the grain from insects.

- Processors may fumigate grain using a modified atmosphere such as carbon dioxide and nitrogen. Fumigation using these inert (nonreactive) gases is popular because chemical use is very restricted. Inert gases do not leave residues, but a storage bin that contains an inert gas can be as deadly to human beings as a lethal chemical, because inert gas does not contain any oxygen.

- Aeration systems can reduce insect feeding and growth in storage bins.

Processors should control dust in handling and storage areas. Keeping dust away saves time spent cleaning floors, walls, ledges, and equipment. Suction (reduced pressure) reduces dust on grain conveyors, receiving hoppers, bucket elevators, bins, and places where grain is transferred from one piece of equipment to another.

Adding very pure oils to grain as it goes into storage can reduce dust. Processors should add the oil (up to 200 ppm) as soon as possible after the grain comes off the truck to reduce dust throughout the plant.

Processors still need to control storage conditions of root crops, such as potatoes, to prevent *Fusarium* tuber rot and bacterial soft rots. The potato industry uses well-ventilated storage rooms with concrete floors to control these problems. Of the potatoes harvested each year, 80 to 90% are stored in good facilities.

Processors usually store bulk oils and shortenings in large carbon or stainless-steel tanks. Staff need to wash these containers with a strong alkaline solution or alkali plus detergents to keep them clean. Nitrogen "blankets" over oils help protect them. Processors need to make sure that oils are not splashed and shaken too much during bottling, because this can cause oxidation, which makes the oil rancid. Processors should coat clean bulk tanks (especially carbon steel tanks) with oil to seal them and prevent rust.

Housekeeping in Raw-Food Storage Areas

Low-moisture food storage areas need good maintenance and housekeeping to keep them sanitary. Processors should maintain bulk storage areas so that they do not have cracks or ledges that can collect dust, dirt, debris, microbes, and insects. Staff should inspect empty bins or other containers to make sure they are in good repair and do not still contain bits of food. Even small amounts of food can allow insects to grow.

Inspectors should look for insect tracks in the dust on floors and walls, and check for moths. They should check for dust in damp areas, where molds, mites, and fungus-feeding insects can grow. Inspectors should also check for unusual smells that could come from mold, insects, or chemicals. It is important to inspect elevators and conveyors that may hold food residues. Equipment that is not being used may hold residues that insects can grow on. Once insects are established in the building, it is easy for them to move in food storage areas.

Staff need to check regularly that storage areas do not have indications of live insects on surfaces, floors, and walls. When processors are going to store grain for a long time, they may use thermocouple cables to monitor the temperature. If the temperature goes up, it may be because insects or molds are growing in the grain. Mold will not grow if the grain is kept dry. If inspectors find insects in the grain, they should treat it or fumigate it. Heat from insects can make moisture spread, which allows mold to grow. Inspectors should keep thorough records of all checks, tests, cleaning, fumigation, or other steps.

Processors should keep storage areas organized so that they can inspect and clean them easily. Well-organized storage areas are less likely to have sanitation problems in the first place. Inspectors and other employees must know how to look for pests and how to get rid of them.

Recommendations for storage include:

- Stacking bags and cartons on pallets
- Spacing bags and cartons away from the walls and from each other for easy inspection and cleaning
- Rotating stock to give insects and rodents less time to get into foods
- Inspecting the area using a flashlight to see into dark corners, under pallets, and between stacks
- Looking for insects that are flying; crawling on walls, ceilings, and floors; and hovering over bags and cartons
- Sifting food to check for insects; (Chap. 9 gives information on controlling insects and rodents.)

Processors should decide how often to clean and inspect storage areas based on the temperature and humidity. At room temperature (25–30°C, 77–86°F), the life cycle of many insects that live in low-moisture grains and foods is about 30 to 35 days. Insects usually stop reproducing below 10°C (50°F). Thus, storage areas should be cleaned and inspected more often when they are warm than when they are cool. The temperature of the raw food affects insects more than the air temperature in the area. Humid areas should be cleaned and inspected more often than dry areas, because moisture allows molds, yeast, and bacteria to grow. Good ventilation and suction can help reduce humidity.

Superheating can kill pests in dry storage areas. But it takes a lot of heat to kill insects, so this process uses a lot of energy, especially during cold weather. Portable heating units can superheat an individual piece of infested equipment. Designs for new or renovated facilities should include equipment for superheating. Keeping storage areas cold enough to stop insects from breeding is not practical, because refrigeration is expensive, and freezing can damage equipment or the facility.

Inspecting Storage Areas

Inspections and written reports of results should be done. The plant should have an inspection report form with a scoring system. The form should describe what each score means. Plants and warehouses may use three rating levels:

- *Acceptable,* if it meets most of the requirements

- *Provisionally acceptable,* if it needs corrective steps to meet the standards
- *Unacceptable,* if it does not meet the standards and the operation is unsanitary

Inspection in processing and storage areas should check for conditions that could contaminate food and correct them before they become a problem. Low-moisture foods have a lower water activity (A_w), so microbes are less of a problem. Other contaminants need more attention in low-moisture-food processing.

Inspectors should check overhead areas for flaking paint, conditions that block cleaning, dust, and condensed water. Inspectors should look for broken window panes and damaged or missing screens at ground level, in the basement, and above ground level. Inspectors should report open windows or other places where pests could get in. They should check for signs of pests, such as insect trails in dust, rodent droppings, and bird droppings or feathers. Employees should report signs of pests so that management can find the source of the problem and correct it. All employees should know how to look for signs of pests.

Operations staff should check the outside of equipment and overhead equipment. Maintenance and sanitation staff should check the inside of equipment to make sure it is sanitary. Some equipment has dead spots where food can collect. Staff need to check these parts regularly when the equipment is not in use. Processors should use equipment with easy access to the inside through cleanout openings or equipment that is easy to take apart. Staff should remove equipment from the area when they are not using it, if possible, and should leave it open so that food particles go though it and can be seen easily.

CLEANING PROCESSING PLANTS

The manufacturing area of low-moisture-food plants should be cleaned every day. Some cleaning should be done while the plant is operating to make sure that it stays tidy. But some cleaning steps and most equipment cleaning (especially the insides) have to wait until production stops. Employees can combine some of the cleaning with routine maintenance. If equipment is easy to reach, employees are more likely to clean it properly and prevent infestation.

Most cleaning equipment is easy to use. Hand brooms, push brooms, and dust and wet mops are the basic cleaning tools. Brushes, brooms, and dustpans remove heavy debris and work well on semismooth floors. Dust mops are faster for cleaning smooth floors with small amounts of dust. Vacuum cleaning is the best way to clean equipment in many production areas. It is thorough and removes light and moderate debris from smooth and rough surfaces. Vacuum cleaners hold dust and do not need a secondary way to collect it (in the way that a broom needs a dustpan). Smaller operations can use portable vacuum cleaners. Larger facilities may need a central (installed) vacuum system. Installed equipment is more expensive but may be more convenient. In large storage areas with nonporous floors, staff may use a mechanical scrubber or sweeper to keep the floor sanitary.

Facilities may have a compressed air line to remove debris from equipment and other difficult-to-reach areas. This is an easy way to clean hard-to-reach areas and is safer than having employees climb ladders with brushes. But compressed air spreads dust from one place over the whole area, so it may spread infestation. Compressed air should have low volume and low pressure so that it does not disperse dust too much. Employees who use compressed air should wear safety equipment such as dust respirators and safety goggles.

Some equipment needs special cleaning tools. Cylindrical brushes are good for spouts. Workers can drag brushes through spouts using rope or cords or use motorized brushes.

Equipment must be well-organized and properly installed so that staff can clean the equipment and the area around it easily. The facility should have specific storage areas for ingredients and supplies. Supervisors should make sure trashcans for bags, film, paper, and waste products are in convenient places in manufacturing, packaging, and shipping areas.

SUMMARY

- Strict sanitation is essential in low-moisture-food-processing and storage facilities. Sanitary facilities make safe food products and meet regulations.
- It is important to choose a good site and a good building design for the plant.
- Processors should choose equipment carefully and design the layout so that the equipment is easy to clean and inspect.
- Samples of raw foods should be collected when they are delivered to make sure they are not infested with pests, molds, rodents, or other contaminants.
- Good housekeeping protects raw and processed foods during storage.
- Storage areas should be inspected often to check for microbes and pests.
- Frequency of inspection and cleaning of storage areas depends on the temperature and humidity.
- The manufacturing area should be cleaned every day.
- In low-moisture product areas, basic cleaning tools can be used, such as vacuum cleaners, powered floor sweepers and scrubbers, and compressed air.

BIBLIOGRAPHY

Foulk, J. D. 1992. Qualification inspection procedure for leased food warehouses. *Dairy, Food and Environ. Sanit.* 12:346.

Marriott, N. G. 1994. *Principles of Food Sanitation,* 3d ed. Chapman & Hall, New York.

Marriott, N. G, Boling, J. W., Bishop, J. R., and Hackney, C. R. 1991. *Quality Assurance Manual for the Food Industry.* Publ. 458–013. Virginia Cooperative Extension, Virginia Polytechnic Institute and State University, Blacksburg.

Mills, R., and Pedersen, J. 1990. *Flour Mill Sanitation Manual.* Eagan Press, St. Paul, Minn.

Troller, J. A. 1993. *Sanitation in Food Processing,* 2d ed. Academic Press, New York.

Walsh, D. E., and Walker, C. E. 1990. Bakery construction design. *Cereal Foods World* 35(5):446.

Worden, G. C. 1987. Freeze-outs for insect control. *AOM Bulletin* (Jan.):4903.

STUDY QUESTIONS

1. What are low-moisture foods?
2. List three things that often spoil low-moisture foods.
3. Why is it important to control dust in low-moisture-food-processing plants?
4. What can processors do to prevent and control infestation of low-moisture foods?
5. What are the advantages and disadvantages of superheating grain?
6. List six types of cleaning equipment used in low-moisture-food-processing plants.

TO FIND OUT MORE ABOUT LOW-MOISTURE-FOOD SANITATION

1. Contact the Pillsbury Company, Pillsbury Center, Minneapolis, MN 55402, and ask for a copy of their leaflet "Preventing or Eliminating Insects in Grain-Based Foods" and other food safety resources.
2. Contact a local bakery or snack food manufacturer, and ask how they keep insects, pests, and mold under control.
3. Put four slices of bread from the same loaf in four plastic bags. Add a tablespoon of water (to add moisture or humidity) to two of the bags. Seal the bags. Place one moist and one dry bag in the refrigerator. Place one moist and one dry bag on the counter. Notice how many days it takes for each slice of bread to show signs of mold. The moist slice of bread at room temperature should grow mold before the others.

C H A P T E R   1 7

Sanitary Food Handling in Foodservice

ABOUT THIS CHAPTER

In this chapter you will learn:

1. Why changes in the foodservice industry have made sanitation even more important
2. Three procedures that can help keep food from being contaminated
3. How clean preparation and serving areas, clean utensils, good employee health and hygiene, sanitary serving areas, and inspection programs can protect food
4. How to clean and sanitize surfaces and equipment in foodservice facilities
5. Why good management and employee training make a sanitation program work properly

INTRODUCTION

The United States has more than 710,000 foodservice establishments, employing about 10 million people. These foodservice operations range from mobile food stands, cafeterias, and fast-food chains to fancy restaurants. Consumers now spend almost half of their food budget on meals away from home. The foodservice industry has grown, and food production, processing, distribution, and preparation have changed. The major changes are more packaged foods, partly or fully prepared bulk foods, individual packages, and food production at a central location.

However much food production, handling, and preparation change, food can still be contaminated with microorganisms and can cause illness. Because of modern processing methods, handling, and distribution, it takes longer for food to reach the table, and it is more likely to be contaminated with microorganisms. Food served in restaurants is responsible for about 58% of outbreaks of foodborne illness. These outbreaks cost somewhere between $1 billion and $10 billion. Centralized kitchens and mass feeding operations mean that more people are affected by a contaminated food. Protecting foodservice customers from foodborne illness is complicated but very important.

The main goal of a foodservice sanitation program is to protect the consumer and prevent or minimize food contamination. It is difficult to protect food from all contamination because pathogenic microorganisms are found almost everywhere and on about half the people who handle food.

SANITARY PROCEDURES FOR FOOD PREPARATION

Prepared food should be kept wholesome and safe by good sanitation during preparation and storage. Three sanitary procedures can help reduce contamination:

1. *Wash food.* Foodservice workers may not need to wash processed foods. However, they should rinse a can before opening it. They should wash all fresh fruits and vegetables before they are prepared (i.e., cut, sliced, peeled), eaten, or cooked, and also wash dried fruits and raisins, raw poultry, fish, and variety meats. Washing poultry reduces contamination of the inside of the poultry. Workers should wash fruits, vegetables, and meats with cold to lukewarm running water to remove dirt, and should drain all washed foods. Soaps and detergents are not used to wash foods because some residue will remain on the food after rinsing. If foods are not cooked immediately, they should be covered and refrigerated.

2. *Protect food.* It is important to protect food from contamination by bacteria that cause foodborne illness. Cleaning compounds, polishes, insect powders, and other compounds used in a foodservice operation can accidentally get into the food. Operators should store all chemicals away from food areas.

3. *Keep food hot or cold.* Processors should thoroughly cook and/or refrigerate all foods that can carry illness-producing microorganisms, including meat, poultry, fish, and dairy products. (In foodservice, most fruits and vegetables are also refrigerated. Cooked food, except some baked goods, is also refrigerated.) Various bacteria need different amounts of heat (different heating times and temperatures) to destroy them and their spores.

REDUCING CONTAMINATION

Preparation and Serving Areas

In food preparation and serving areas, air, surfaces, and people touch the food, and it is open to contamination with dirt and microorganisms that cause food poisoning and food spoilage. So it is very important to keep these areas clean and to cover the food.

Utensils

Staff should thoroughly wash and disinfect utensils to keep them hygienic and stop them from spreading bacteria between foods. After cleaning, utensils should be sanitized by heating to 77°C (171°F) for at least 30 seconds or by using chemical germicides at room temperature for at least 10 minutes (see Chap. 6 for more information about sanitizing). Cracked, chipped, or dented dishes or utensils can carry bacteria—and should be thrown away. Food and microorganisms can collect in cracks and dents, and cleaning and sanitizing may not remove them.

Servers should not touch any surface that will touch customers' mouths or the food. Microorganisms can transfer from hands and surfaces to dishes, utensils, or food, and then to consumers.

Employees' Health

Foodservice operations should not hire employees who have or are carriers of any disease that could be carried by food, water, or utensils. If workers have colds, flu, or skin infections, supervisors should give them responsibilities where they do not come in contact with food, or supervisors should not let them work until they are healthy.

Many local health codes require health checks every 6 months or 1 year. Employees may need a health card stating that they do not have an infectious disease. These examinations do not protect consumers against the short-term diseases that most workers experience every year, so it is important to train employees to report infections and visit their doctor when necessary. Training should also include personal hygiene to reduce the spread of disease (see Chap. 4).

Buffet or Cafeteria Service

The following can reduce contamination during buffet or cafeteria service:

- Cooling tables should keep foods at 2°C (36°F) or cooler so that microorganisms cannot grow quickly.
- Warming tables should keep foods at 60°C (140°F) or hotter.
- Authorized workers should serve food in hygienic conditions; customers should not handle serving utensils.
- Servers should wear hairnets or hats, and should not touch their hair while they are serving so that they do not contaminate the food.
- Operators should place transparent shields between the customers and the service tables and dishes so that customers cannot handle foods, or breathe, cough, or sneeze on foods.

Inspection Programs

The foodservice establishment should encourage local public-health authorities to inspect the facilities. If the local health department uses checksheets, the foodservice establishment should use a copy of these sheets for its own sanitation program. The operator should keep records of inspection reports and how the facility corrected any problems.

The Food and Drug Administration is promoting the Hazard Analysis Critical Control Points (HACCP) concept to stress factors that are known to cause foodborne disease. The National Restaurant Association (NRA) introduced a similar self-inspection tool, the Sanitary Assessment of Food Environment (SAFE), to the foodservice industry. Both tools have the same goal of consumer protection and stress safe food handling to reduce contamination and prevent pathogens from growing in foods.

County health departments throughout the United States are using the SAFE concept to inspect restaurants. SAFE looks at control of time and temperature at each step during preparation of foods that could be hazardous. (Foods should be kept hot or cold, and the duration of time the food is between 4°C (40°F) and 60°C (140°F) should be as short as possible (see Fig. 12.1.) SAFE protects foodservice establishments from outbreaks of foodborne disease by identifying, controlling, and monitoring critical points in the food-handling and preparation operation. This inspection program spends the most effort on foods that have the highest risk and on the production steps where the food is most likely to be contaminated. When operators have identified the critical control points, they can plan production and sanitation to make sure the food is safe.

SAFE is a simpler version of HACCP, and is based on three principles:

1. Keep it clean.
2. Keep it cold.
3. Keep it hot.

SAFE makes the foodservice establishment less likely to be held liable for outbreaks of foodborne illness because it can show that it has done all it could to keep the food safe. The first step in designing a SAFE program is to diagram each preparation step and record food temperatures every 30 minutes. A SAFE program should concentrate on moist, nonacid, high-protein foods, such as eggs, milk and dairy products, most meats, poultry, seafood, sauces and gravies, cooked cereal grains, and cooked vegetables.

A SAFE program should look for processing steps that tend to increase the risk of foodborne disease:

Multiple preparation steps: More handling means more chances to contaminate the food.

Temperature changes. Foods are at temperatures in the danger zone (4–60°C, 40–140°F) during heating and cooling.

Large volume: Large volumes of food take longer to heat and cool, so microorganisms have more time to grow.

Contaminated basic foods: Field dirt or pesticides can contaminate raw produce; raw red meats and poultry can be contaminated during slaughter; and raw seafood can carry viruses, bacteria, or parasites.

The SAFE program checks food items from the time they arrive at the receiving area until workers serve them to customers. Workers record temperatures and times at the beginning and end of each handling step. A SAFE survey identifies several control points, but only a few are critical control points. These include:

• Establishing good personal hygiene procedures and keeping equipment, utensils, and surfaces clean

- Holding foods at 2°C (36°F) or lower or 60°C (140°F) or higher
- Destroying microbes by cooking foods to an internal temperature hotter than 74°C (165°F)

Critical controls and monitoring points should be realistic. If a step does not meet its critical control goal (if the temperature is not cold enough or hot enough, if the temperature is in the danger zone for too long, or if cleaning and sanitation are not done properly), management may need to redesign the process.

CLEANING AND SANITIZING

The foodservice operator can choose from many cleaning and sanitizing methods and products. The operator needs to decide which methods and products will work best and must be certain that they are used properly.

Cleaning Basics

Foodservice establishments need a planned cleaning and sanitizing schedule and proper supervision. If workers are hurrying to meet the needs of customers, they often neglect parts of the sanitation program. The manager must understand food safety, be alert, and insist on disciplined sanitation. He or she must know how to be certain that conditions are sanitary. Workers must clean and sanitize all surfaces that touch food every time they are used, when there is a break in service, or at set times if employees use the surfaces all the time.

Cleaning uses the principles of chemistry in a practical way. Sanitarians should choose the best cleaning compound for each job. A compound that works well for one cleaning job may not be satisfactory for another cleaning job. See Chapter 5 for more information about cleaning compounds. Cheap cleaning compounds are not always more economical than those that are more expensive, because workers may need to use more of the cheaper compounds if they do not work as well as those that are expensive.

Alkaline cleaners do not work on some types of soil—for example, lime crusts on dishwashing machines, rust stains in washrooms, and tarnish on copper and brass. Acid cleaners, usually with a detergent, work well to remove these soils.

If soil is fixed so firmly to a surface that alkaline or acidic cleaners cannot remove it, a cleaner containing a scouring agent (usually finely ground feldspar or silica) usually works. Abrasives work well on worn and pitted porcelain, rusty metals, or very dirty floors. Workers should be careful with use of abrasives in a foodservice facility because they can scratch smooth surfaces.

Sanitizing Basics

Operators use heat or chemicals to sanitize equipment, surfaces, and utensils. Neither type of sanitizing works if staff do not first clean and rinse equipment, surfaces, and utensils properly. Any soil that stays on the surface can protect microorganisms from the sanitizer. (See Chap. 6 for more information about sanitizing.)

Because cooking utensils are heated during cooking, they may not seem to need to be sanitized. But heat from cooking does not always make every part of the utensil hot enough for long enough to make sure it is sanitized. During heat sanitizing, the temperature must be hot enough for long enough to kill microorganisms.

During chemical sanitizing, the clean object is immersed in the sanitizer for about 1 minute, or the surface is rinsed or sprayed with sanitizer at twice the normal strength. The strength of the sanitizing solution should be tested frequently, because the sanitizer is used up as it kills bacteria. The sanitizer should be changed when it is not strong enough to work properly. Firms that make sanitizers normally provide free test kits to check the strength of the sanitizer. Workers should only use toxic sanitizing agents on surfaces that do not touch the food.

Detergent-sanitizers contain a sanitizer and a cleaning compound. These products can sanitize, but sanitizing should be done in a separate step from cleaning because the sanitizing power can be lost during cleaning if the chemical sanitizer reacts with the soil. Detergent-sanitizers tend to be more expensive than plain cleaning compounds and have limited uses.

Employees can clean and sanitize most portable items that touch food at a washing area away from the food preparation area. The work station should have three or more sinks, separate drain boards for clean and soiled items, and an area for scraping and rinsing food waste into a garbage container or disposal. If the sanitizing step uses hot water, the third compartment of the sink must have a heating unit and a thermometer to make sure the water stays at about 77°C (171°F). Different areas in the facility may need to meet different regulations for cleaning and sanitizing.

Cleaning Steps

Manual cleaning and sanitizing of a typical foodservice facility has eight steps:

1. Clean sinks and work surfaces before each use.
2. Scrape off heavy soil and presoak to remove large amounts of soil that could use up the cleaning compound. Sort items and presoak silverware and other utensils in a special soaking solution.
3. Wash items in the first sink in a clean detergent solution at about 50°C (122°F), using a brush or dish mop to remove the soil.
4. Rinse items in the second sink. This sink should contain clear, clean water at about 50°C (122°F). This removes traces of soil and cleaning compounds that may stop the sanitizing agent from working properly.
5. Sanitize utensils in the third sink by dipping them in hot water (82°C, 180°F) for 30 seconds or in a chemical sanitizing solution at 40 to 50°C (104–112°F) for 1 minute. The sanitizing solution should be mixed at twice the recommended strength so that it does not get too dilute when items carry water from the rinse sink. It is important to get rid of air bubbles that could stop the sanitizer from getting inside items.
6. Air-dry sanitized utensils and equipment. Wiping can recontaminate them.

7. Store clean utensils and equipment in a clean area more than 20 cm (8 in) off the floor to protect them from splashing, dust, and contact with food.
8. Cover fixed equipment when not using it, especially the parts that touch the food.

Stationary Equipment

Food preparation equipment that cannot be moved should be cleaned according to the manufacturer's instructions for taking it apart and cleaning it. Here are some general steps:

1. Unplug all electrical equipment.
2. Take all equipment apart; wash and sanitize every part.
3. Wash and rinse all surfaces that touch food with a sanitizing solution at twice the strength used to sanitize by dipping.
4. Wipe all surfaces that do not touch food. Dip cloths used to wipe down equipment and surfaces in a sanitizing solution. Keep these cloths separate from other wiping cloths.
5. Air-dry all clean parts before putting them back together.
6. Follow the manufacturer's instructions for equipment that is designed to have detergent and sanitizing solutions pumped through it. Use high-pressure, low-volume cleaning equipment (see Chap. 7) and spray devices for sanitizing. To sanitize, spray with double-strength solution of sanitizer for 2 to 3 minutes.
7. Scrub wooden cutting boards with a nontoxic detergent solution and stiff nylon brush (or use a high-pressure, low-volume cleaning wand). Apply a sanitizing solution after every use. Replace worn, cut, or scarred wooden boards with polyethylene boards. Do not dip wooden cutting boards in a sanitizing solution.

Floor Drains

Workers must clean floor drains every day after cleaning everything else. Sanitation workers should wear heavy-duty rubber gloves to remove the drain cover and should use a drain brush to remove the debris. When they replace the cover, they should flush it with a hose going through the drain so that it does not splatter. Workers should pour a heavy-duty alkaline cleaner at the strength the manufacturer recommends down the drain, wash the drain with a hose or drain brush, and rinse it. Workers should use a quaternary ammonium plug (Quat Plug), or pour a chlorine or quat sanitizing solution down the drain.

Cleaning Tools

Foodservice operators should store cleaning tools separately from sanitizing tools. They should rinse, sanitize, and air-dry cloths, scrubbing pads, brushes, mops, and sponges after use. Clothing should be laundered every day. All buckets and mop pails should be emptied, washed, rinsed, and sanitized at least once a day.

Mechanized Cleaning and Sanitizing

Mechanized cleaning can work better than hand cleaning if it is used properly. More operations are beginning to understand the importance of good sanitation, and as operators prepare greater volumes of food, more of them are using dishwashing machines. Larger foodservice establishments may also use portable high-pressure, low-volume cleaning equipment.

There are two basic types of dishwashing machine: high-temperature washers and chemical sanitizing machines.

High-temperature washers. The four main types of high-temperature washer are discussed below. The sanitizing temperature for these washers should be at least 82°C (180°F), but not higher than 90°C (194°F).

1. *Single-tank, stationary-rack-type with doors.* This washer contains racks that do not move during the wash cycle. Utensils are washed using a cleaning compound and water at 62 to 65°C (144–149°F) sprayed from underneath the rack, but the machine may also have sprayers above the rack. After the wash cycle, the machine has a hot-water final rinse.

2. *Conveyor washer.* This equipment has a moving conveyor that takes utensils through the washing (70–72°C, 158–162°F), rinsing, and sanitizing (82–90°C, 180–194°F) cycles. Conveyor washers may contain a single tank (each step in the cycle is done one at a time) or multiple tanks (one each for washing, rinsing, and sanitizing).

3. *Flight-type washer.* This washer is a high-capacity, multiple-tank unit with a peg-type conveyor. It may have a built-on dryer and is common in large foodservice facilities.

4. *Carousel or circular conveyor washer.* This multiple-tank washer moves a rack of dishes on a peg-type conveyor or in racks. Some models stop after the final rinse.

Chemical sanitizing washers. The three main types of chemical sanitizing dishwasher are discussed below. Glassware washers are another type of chemical sanitizing machine. The water temperature during chemical sanitizing should be 49 to 55°C (120–131°F).

1. *Batch-type dump.* This washer washes and rinses in one tank. The machine times each step and releases the cleaning compound and sanitizer automatically.

2. *Recirculatory door-type, nondump washer.* The water does not completely drain from this washer between cycles. The machine dilutes the water with freshwater and reuses it during the next cycle.

3. *Conveyor-type, with or without a power prerinse.* This machine uses a conveyor to carry utensils and dishes from one cycle to the next and may or may not include a power prerinse.

When buying or using dishwashing equipment, it is important to consider the following:

1. Be certain that the dishwasher is the right size to handle the load.
2. If the equipment uses hot water to sanitize, it will need a booster heater that can provide water at 82°C (180°F).
3. Thermometers must be accurate to make sure that the water is at the right temperature.
4. It is important to install, maintain, and operate the equipment properly so that it cleans and sanitizes properly.
5. The layout of the dishwashing area should be convenient and efficient for staff and for the operation.
6. Operators should consider whether they need a prewash cycle instead of scraping and soaking very dirty utensils.
7. In machines with compartments, rinse-water tanks should be protected so that washwater cannot contaminate rinse water.
8. Staff should clean large dishwashers at least once a day following the manufacturer's instructions.

   Table 17.1 gives some ideas for troubleshooting problems with dishwashers.

Cleaning-in-place (CIP) equipment. Equipment such as automatic ice-making machines and soft-serve ice-cream and frozen yogurt dispensers are designed so that staff can clean them by running a detergent solution, hot-water rinse, and sanitizing solution through the unit. Operators should look for machines that keep the cleaning and chemical sanitizing solution in a fixed system of tubes and pipes for a set amount of time. The cleaning water and solution should not be able to leak into the rest of the machine. The cleaning and sanitizing solutions must touch all surfaces that touch the food; the CIP equipment must drain itself; and operators must be able to look inside to make sure that the unit is clean.

Recommendations for Cleaning Specific Areas and Equipment

Area. Floors

Frequency. Daily

Supplies and equipment. Broom, dustpan, cleaning compound, water, mop, bucket, and powered scrubber (optional)

Daily

1. Clean all table surfaces (wipe pieces of food into a container, wash table with warm soapy water, rinse surfaces with clean water).
2. Stack chairs on tables or take them out of the area.
3. Sweep and remove trash using a push broom.
4. Put up signs to warn people that the floor is wet.
5. Mop floors or use a mechanical scrubber in larger operations. Use detergent (15 g/L or 2 oz/gal) and rinse with clear water.

TABLE 17.1.   Dishwashing Problems and Solutions

Problem	Possible Cause	Suggested Solution
Dirty dishes	Not enough detergent	Use enough detergent in washwater to remove and suspend soil.
	Washwater too cool	Keep water at recommended temperature to dissolve food, and preheat for sanitizing step.
	Wash and rinse times too short	Allow enough time for wash and rinse steps to work (use an automatic timer or adjust conveyer speed).
	Bad cleaning	Unclog rinse and wash nozzles to keep the right pressure, spray pattern, and flow. Make sure overflow is open. Prescrape dishes to keep washwater as clean as possible. Change water in tanks more often.
	Bad racking	Make sure that racks are loaded properly, based on the size and type of utensils. Presoak silverware and place in silver holders without sorting or shielding.
Films	Hard water	Use a water softener outside the machine. Use a detergent that contains a water conditioner. Make sure that washwater and rinse water are not too hot (hot water may cause film to settle out).
	Detergent film	Make sure that there is enough rinse water at high-enough pressure. Make sure that wash jets are not worn or spraying at the wrong angle and getting into rinse water.
	Poor cleaning or rinsing	Clean the machine often and carefully so that scale does not build up inside it. Make sure that there is enough water with enough pressure.
Greasy films	Too acidic (low pH); not enough detergent; water too cool; equipment not cleaned properly	Make sure that cleaning solution is alkaline enough to suspend grease. Check cleaning compounds and water temperatures. Unclog wash and rinse nozzles so they can spray properly. Change water in tanks more often.
Foam	Detergent type or solids suspended in water	Use a low-sudsing detergent. Reduce the amount of solids in the water.

Table 17.1 (*Continued*)

Problem	Possible Cause	Suggested Solution
Foam (*continued*)	Very dirty dishes	Scrape and prewash dishes so that food does not decompose and cause foam.
	Equipment not cleaned properly	Make sure that spray and rinse nozzles are open. Keep equipment free of deposits and films that can cause foam.
Streaks	Alkaline water, solids dissolved in water	Treat the water to make it less alkaline. Use a rinse additive. Treat the water outside the machine if the problem is severe.
	Bad cleaning or rinsing	Make sure that there is enough rinse water at high-enough pressure. Rinse alkaline cleaners off dishes thoroughly.
Spots	Hard rinse water	Soften the water inside or outside the machine. Use rinse additives.
	Rinse water too hot or too cold	Check rinse-water temperature. Dishes may be flash drying, or water may be drying on dishes rather than draining off.
	Not enough time between rinsing and storage	Change to a low-sudsing cleaner. Treat the water to reduce the solid content.
	Food soil	Remove most of the soil before washing. Change water in the tanks more often.
Stains: coffee, tea, or metal	Wrong detergent	Use a detergent with chlorine.
	Equipment not cleaned properly	Make sure spray and rinse nozzles are open. Keep equipment free of deposits and films.

Source: National Sanitation Foundation (1982).

Weekly

1. Do the daily cleaning.
2. Scrub floors (use a powered scrubber and/or buffer on floor, rinse with clean water at 40–55°C [104–131°F] water. Sponge-mop and dry-mop the floor).

Area. Walls

Frequency. Daily and weekly

Supplies and equipment. Handbrush, sponge, cleaning compound, bucket, water, and scouring powder

Daily

1. Clean spots that look dirty. Use cleaning compound (15 g/L or 2 oz/gal), hand-wipe dirty areas, rinse with clean water, wipe dry.

Weekly

1. Remove debris from walls.
2. Mix 15 g of cleaning compound per liter of water (2 oz/gal).
3. Scrape walls using a handbrush. Scrub tiles and grout.
4. Rinse wall surfaces with clean, warm water.
5. Wipe dry using clean cloths or paper towels.

Area. Shelves

Frequency. Weekly

Supplies and equipment. Handbrush, detergent, sponge, water, and bucket

1. Remove items from the shelves, and store on a pallet or other shelves.
2. Brush off all debris into a pan or container.
3. Clean shelves in sections (scrub with detergent and warm water).
4. Replace items on the shelves (throw away damaged goods).
5. Mop floor area under shelves with a clean, damp mop.

Equipment. Stack oven

Frequency. Clean once a week thoroughly; wipe daily

Supplies and equipment. Salt, metal scraper with long handle, metal sponges, cleaning compound in warm water, 4-liter (1-gal) bucket, sponges, stainless-steel polish, ammonia, vinegar, or oven cleaner if appropriate

Weekly

1. Turn off the heat and scrape the inside. Sprinkle salt on hardened oven spills. Turn the oven on at 260°C (500°F). When the spills have completely carbonized (blackened), turn off the oven. Cool thoroughly. Scrape the floor with a long-handled metal scraper. Use a metal sponge or hand scraper on the inside of doors, handles, and edges.
2. Brush out scraped carbon and other debris. Begin with top deck of stack oven. Brush out with stiff bristle brush, and use dustpan to collect debris.
3. Wash doors (use a hot detergent solution only on enamel surfaces, rinse, and wipe dry).
4. Brush inside using a small broom or brush.
5. Clean and polish outside of oven (wash the top, back, hinges, and feet with warm cleaning compound solution, rinse, and wipe dry; polish all stainless steel).

Note: Never pour water on the outside of the oven, and do not use a wet cloth or sponge on the outside surfaces. Do not sponge, drip, or pour water inside the oven to clean it.

Equipment. Hoods

Frequency. Weekly (at least)

Supplies and equipment. Rags, warm soapy water, stainless-steel polish, degreaser for filters

1. Remove filter, carry it outside, rinse with a degreaser, and run it through the dishwasher after all dishes and eating utensils have been cleaned.
2. Wash inside and outside of hood (use warm, soapy water and a rag to wash hoods completely on the inside and outside to remove grease; clean drip trough in area below filters).
3. Shine hood with polish (spray polish on hood and wipe off; use a clean rag on inside and outside).
4. Replace filters.

Equipment. Range surface unit

Frequency. Weekly

Supplies and equipment. Putty knife, wire brush, damp cloth, hot detergent solution, 4-liter (1-gal) container, vinegar or ammonia as appropriate

1. To clean back apron and warming oven or shelf, let surfaces cool. Use a hot, damp cloth wrung almost dry. Remove hardened dirt with a putty knife, and scrape edge of plates. Scrape burned material from top surfaces with a wire brush.
2. Lift plates. Remove burned particles with a putty knife, and scrape edge of plates. Scrape burned material from flat surfaces with a wire brush.
3. Wipe heating elements with a damp cloth.
4. Clean base and exterior (wipe with a cloth dampened with hot detergent water).
5. Clean grease receptacles and drip pans (soak grease receptacles and drip pans in a detergent solution for 20 to 30 min, scrub, rinse, and dry).

   Note: Do not dip heating elements in water.

Equipment. Griddles

Frequency. Daily

Supplies and equipment. Spatula, pumice stone, paper towels, hot detergent solution

1. Turn off heat. Remove grease after each use. When surface is cool, scrape it with a spatula or pancake turner. Wipe clean with dry paper towels. Use a pumice stone block to clean burned areas on plates. (Do not use pumice stone daily.)
2. Clean grease and drain troughs (pour a hot detergent solution into a small drain and brush; rinse with hot water).

3. Empty grease receptacles, and remove grease with hot detergent solution; rinse and dry.

4. Scrub guards, front, and sides of the griddle using a hot detergent solution. Wash off grease, splatter, and film. Rinse and dry.

Equipment. Rotary toaster

Frequency. Daily

Supplies and equipment. Warm detergent solution, brush, cloths, stainless-steel polish, nonabrasive cleaner

1. After the toaster is cooled, disconnect it and take it apart (remove pan, slide, and baskets).

2. Clean surface and underneath using a soft brush to remove crumbs from the front surface and behind bread racks.

3. Clean frame and inside as far as you can reach. Wipe clean with a warm hand-detergent solution. Rinse and dry. Polish if necessary with a nonabrasive cleaning powder. The outside casing should not collect grease or dirt. Do not let water and cleaning compounds touch the conveyor chains. Polish the frame if it is made of stainless steel.

Equipment. Coffee urns

Frequency. Daily

Supplies and equipment. Outside cleaning compound (stainless-steel polish), inside cleaning compound (baking soda), urn brush, faucet, and glass brush

1. Rinse urns by flushing with cold water.

2. Be certain that the outer jacket is three-quarters full of water. Turn on heat. Open water inlet valve and fill coffee tank with hot water to the coffee line. Add recommended quantity of cleaning compound (10–15 g/L, 1.3–2.0 oz/gal). Allow solution to remain in the liner for approximately 30 min with the heat on full.

3. Brush the liners, faucet, gauge glass, and draw-off pipe (scrub inside of tank, top rim, and lid); draw off 2 L (1/2 gal) of solution, and pour it back to fill the valve and sight gauge. Put the brush in the gauge glass and coffee draw-off pipe, and brush briskly.

4. Drain by opening the coffee faucet and completely draining the solution. Close the faucet.

5. To rinse the machine, open the water inlet valve into the coffee tank. Use 4 L (1 gal) of hot water. Open the faucet for 1 min to allow water to flow and sterilize the dispenser

6. Take faucet apart and thoroughly scrub with a brush. Rinse spigot thoroughly.

7. Twice weekly, make a solution of 1 cup baking soda in 4 L (1 gal) hot water, and hold in the urn for approximately 15 min. Drain. Flush thoroughly with hot water before use.

Note: Put a tag on the faucet while the urn is soaking with the cleaning compound.

Biweekly

1. Fill urn with a destaining compound solution (2 tablespoons to 20 L [5 gal] of water at 80°C [176°F] or as directed by manufacturer).
2. Open spigot and draw off 4 L (1 gal) of solution. Pour this back into the tank, and let it stand for 1 hr at 75 to 80°C (167–176°C).
3. Scrub liner and gauge glass using a long-handled brush to loosen scale.
4. Clean faucet (take faucet valve apart and clean each piece; soak in hot water).
5. Rinse valve parts and inside of urn 3 or 4 times with hot water until all traces of cleaning compound are removed; put back together.
6. Refill urn.

Note: To remove stain from vacuum-type coffeemakers, use a solution of 1 teaspoon of compound per liter (quart) of warm water. Fill the lower bowl up to within 5 cm (2 in) of the top, and put the unit together.

Equipment. Iced-tea dispensers

Frequency. Daily

Supplies and equipment. Rags and warm, soapy water

1. Clean outside with a damp cloth.
2. Empty drip pan and wash drip pan and grill with a mild detergent and warm water.
3. Wash trough (open front jacket, remove mix trough, and wash in a mild detergent and warm water).
4. Wash plastic parts (do not soak plastic parts in hot water or wash in dishwashing machines).

Equipment. Steam Tables

Frequency. Daily

Supplies and equipment. Dishwashing detergent, spatula, scrub brush, and cloths

1. Turn off heating unit by turning steam valve counterclockwise for steam-heated food or turning dial to OFF position for electrically heated food.
2. Remove insert pans and take them to the dishwashing area after each use. Hand-clean, sanitize, and air-dry. Store in a clean area until needed.
3. Drain water from the steam table. Remove the overflow pipe, using a cloth to prevent injury.
4. Prepare the cleaning solution (30 ml [1 fl oz] of dishwashing liquid in a suitable container).

5. Scrape out food particles from the steam table with a spatula or dough scraper.
6. Scrub inside, and clean outside with a scrub brush and cleaning solution.
7. Rinse outside using enough clear water to remove all of the detergent.

Note: Mobile electric hot-food tables: Clean corrosion-resistant steel after each use. Remove ordinary deposits of grease and dirt with mild detergent and water. Rinse thoroughly and dry.

Equipment. Refrigerated salad bars (with ice beds or electrically refrigerated)

Frequency. After each use

Supplies and equipment. Detergent, plastic brush, and sanitizing agent

1. Transfer shallow pans or trays and containers to preparation areas after each meal. Run insert pans and/or trays through dishwashing machine.
2. Clean table counter using detergent and plastic brush. Rinse and sanitize by swabbing with a solution containing a sanitizing agent.
3. Descale whenever needed to prevent rust, lime, or hard-water deposits in those with ice beds. Fill table bed with boiling water, and add a descaling compound following the manufacturer's recommendations. Allow to stand for several hours. Scrub with a plastic brush. Drain. Rinse thoroughly. Sanitize by spraying with a sanitizing solution.
4. Defrost electrically refrigerated units as often as needed. Follow up with a cleaning procedure as described above.

Equipment. Milk dispenser

Frequency. Daily

Supplies and equipment. Sanitized cloth or sponge, mild detergent, sanitizing agent

1. To remove empty cans from the dispenser, place a container under the valve, open the valve, and tip the can forward in the dispenser to drain out the remaining milk. Remove the tube and lift out the oar.
2. Wipe up any spills using a sanitized cloth or sponge.
3. Wash the whole inside surface with a milk cleaning solution and rinse.
4. Clean the outside using a stainless-steel cleaner. If the steel is discolored or stained, swab with a standard chemical, leave for 15 to 20 min, rinse with clear water, and polish with a soft cloth.
5. Take apart and clean the valves every day or each time empty cans are removed. Wash in detergent water. Rinse and sanitize.
6. Wipe the bottom of full milk cans with a sanitizing solution before placing in the dispenser.

Equipment. Deep fat fryer

Frequency. Daily

Supplies and equipment. Knife, spatula, wire brush, detergent, long-handled brush, vinegar, nylon brush, dishwashing compound

Daily

1. Turn off the heating element. Allow fat to cool to 65°C (149°F).
2. Drain and filter the fat after each use. Open drain valve, drain entire kettle contents, and filter into a container. Remove fat container. Place a clean fat container into the well or wash, and replace the original container.
3. Remove baskets, and scrape off the oxidized fat with a knife. Remove loose food particles from the heating units with a spatula or a wire brush. Flush down sides of the kettle with a scoop of hot fat. Soak basket and cover in a deep sink with hot detergent.
4. Close the drain. Fill the tank with water up to the fat level; add 60 ml (2 fl oz) of dishwashing compound.
5. Turn on heat to 121°C (250°F) and boil 10 to 20 min.
6. Turn off heat, open drain, and draw off cleaning solution.
7. Scrub inside using a long-handled brush. Flush out with water. Clean the basket with a nylon brush, and place it back in the kettle.
8. Fill the kettle with water to rinse. Add 1/2 cup of vinegar to neutralize the remaining detergent. Turn on the heating element. Boil 5 min to sanitize and *turn off heat.* Drain. Rinse with clear water.
9. Air dry all parts, including baskets and strainer.
10. When kettle is cool, wipe off the outside with a grease solvent or a detergent solution. Rinse.

Weekly

1. Fill kettle with water to fat level. Heat to at least 80°C (176°F) or boil for 5 to 10 min. Turn off the heat.
2. Add 1/2 tablespoon of destaining compound (stain remover for tableware) per liter (quart) of water. Stir solution, and loosen particles on sides of kettle.
3. Place screens and strainers in water at 80°C (176°F) containing 1/2 tablespoon of destaining compound per liter (quart). Allow to stand overnight. Rinse thoroughly and air-dry.
4. Drain kettle and rinse thoroughly before replacing cleaned screen and strainer.

Equipment. Vegetable chopper

Frequency. Daily

Supplies and equipment. Brush, sponge, cloth, bucket, detergent, sanitizer solution

1. Take parts apart after each use (turn off power first, and wait until knives have stopped revolving).

2. Clean knives, bowl guard, and bowl. Remove blades from the motor shaft and clean carefully. Wash with a hand detergent solution. Rinse and air-dry. Remove all food particles from the bowl guard. If the bowl is removable, wash it with other parts; if the bowl is fixed, wipe out food particles from table or base. Clean with a hand detergent solution; rinse and air-dry.

3. Clean parts and under chopper surface. Dip small parts in a hot hand detergent solution; wash, rinse, and air-dry.

4. Put the parts back together. Replace comb in guard. Attach bowl to the base and knife blades to the shaft. Drop guard into position.

   Note: The mechanical details of choppers vary a lot.

Equipment. Meat slicer

Frequency. Daily

Supplies and equipment. Bucket, sponge, cloth, brush, detergent, sanitizer solution

1. Disconnect the equipment. Remove meat holder and chute by loosening screw. Remove scrap tray by pulling it away from the knife. Remove the knife guard. Loosen bolt at the top of knife guard in front of the sharpener. Remove bolt at the bottom of the knife guard behind chute. Remove guard.

2. Clean slicer parts by scrubbing in a sink filled with hot detergent solution. Rinse with hot water. Immerse in a sanitizer solution. Air-dry.

3. Wipe the knife blade with a hot detergent solution. Wipe from center to edge. Air-dry.

4. You may use a sanitizer block to sanitize the blade. This block is sliced by the blade after the slicer is put back together.

5. Clean receiving tray and underneath tray with a hot detergent solution. Rinse in hot clear water. Air-dry.

   Note: Do not pour water on equipment or immerse this equipment in water.

SANITATION MANAGEMENT AND TRAINING

Proper cleaning does not happen without good management. For a good cleaning program, management must make an effort to understand the type of cleaning the food-service establishment needs and to provide the right equipment and chemicals for the jobs. A good cleaning program makes sure that cleaning jobs are not forgotten, forces the sanitation manager to plan ahead and make the best use of resources, helps new employees with cleaning routines, gives a basis for inspections, and saves employees from having to decide what to clean when. Table 17.2 shows part of a cleaning schedule for a food preparation area. A full cleaning schedule could use the same format. Management should arrange the schedule logically so that nothing is forgotten.

Management should schedule most cleanups when they are least likely to contaminate foods or interfere with service. Staff should not vacuum and mop while food is being prepared and served, but they should clean as soon as possible afterwards so that soil does not dry and harden and bacteria do not have a chance to grow.

TABLE 17.2. Part of a Sample Cleaning Schedule for Food Preparation Area

Item	When	What	Use	Who
Floors	As soon as possible	Wipe up spills	Cloth, mop and bucket, broom and dustpan	———
	Once per shift between rushes	Damp mop	Mop, bucket, or mechanical scrubber	———
	Weekly, Thursday evening	Scrub	Brushes, bucket, detergent	———
	January, June	Strip, reseal	See procedure	———
Walls and ceilings	As soon as possible	Wipe up splashes	Clean cloth; portable high-pressure, low-volume cleaner; or portable foam cleaner	———
	February, August	Wash walls		Contracted specialists
Work tables	Between uses and at end of day	Clean and sanitize tops	See cleaning procedure for each table	———
	Weekly Saturday	Empty, clean, and sanitize drawers; clean frame, shelf	See cleaning procedure for each table	———
Hoods and filters	When necessary	Empty grease traps	Container for grease	———
	Daily, closing	Clean inside and out	See cleaning procedure	———
	Weekly, Wednesday evening	Clean filters	Dishwashing machine	———
Broiler	When necessary	Empty drip pan, wipe down	Container for grease; clean cloth	———
	After each use	Clean grid tray, inside, outside, top	See cleaning procedure for each broiler	———

Adapted from *Applied Foodservice Sanitation*, 4th ed. Copyright 1992 by the Educational Foundation of the National Restaurant Association.

Managers should arrange the schedule so that cleaning times are evenly spaced and tasks are done in the right order.

Managers should discuss new cleaning programs with employees at a meeting and should demonstrate use of new equipment and procedures. It is important to explain why the program is being changed, how it will benefit the operation, and why employees should follow procedures exactly.

Managers should evaluate the sanitation program to make sure it is working properly. They may evaluate the program during on-going supervision and inspections. They must monitor the program to make sure that each step is followed. Managers should keep written records to show that employees are following the program.

Employee Training

Training takes workers and management away from the job. To make it worthwhile, many operations employ training specialists. Printed sheets, posters, demonstrations, slides, and films are all helpful in training.

It is difficult to prove that sanitation training is cost-effective. The operation cannot always measure the benefits. But sanitation could prevent a costly outbreak of foodborne illness or save an establishment from having to close until local health standards are met. It also improves the image of the operation, which tends to increase sales.

Employee training is important because it is difficult to hire capable and motivated workers. Training should be ongoing because the foodservice industry has a high rate of employee turnover, and employees need frequent reminders to keep good sanitation on their minds.

On-the-job training can work for some tasks but is not complete enough for sanitation training. Each employee involved in foodservice should understand what sanitation means and what he or she needs to do in his or her job to keep conditions sanitary.

The best way to train employees in a large or medium-sized firm is to set up a training department and hire a training director. Foodservice trainers have established their own professional association, the Council of Hotel and Restaurant Trainees.

In most small foodservice operations, supervisors are responsible for sanitation training. A trained employee or someone certified in foodservice sanitation should do the training.

Management can tell how well a training program works by whether employees can follow the sanitation program. The number of guest complaints and customer return rates also show if the training program works.

The National Restaurant Association Educational Foundation (1992) recommends two methods to evaluate training. An objective method uses tests or quizzes to find how much employees understand. In the other method, managers evaluate employees' job performance. Training works better when managers praise employees and use wall charts, pins, and certificates for good performance. Organizations such as the National Restaurant Association Educational Foundation and some regulatory agencies have certification courses that give training and recognition.

SUMMARY

- Today, food is handled more, so it is harder to protect it from contamination.
- Hygienic design of the facility and equipment makes cleaning easier in foodservice establishments.
- Several organizations and manufacturers have equipment standards that make it easier to choose sanitary equipment.
- The facility, equipment, and utensils must be thoroughly cleaned and sanitized.
- Mechanized cleaning using a dishwasher is a good way to clean and sanitize equipment and utensils, provided the dishwasher is well maintained and properly operated.
- Managers need a written cleaning and sanitizing program . They must train, supervise, and evaluate how well employees carry out the program.

BIBLIOGRAPHY

Anon. 1982. News and views. *National Provisioner* 186:2.

Guthrie, R. K. 1988. *Food Sanitation,* 3d ed. Chapman & Hall, New York.

Harrington, R. E. 1987. How to implement a SAFE program. *Dairy and Food Sanitation* 7:357.

Longree, K., and Blaker, G. G. 1982. *Sanitary Techniques in Food Service,* 2d ed. John Wiley, New York.

Marriott, N. G. 1994. *Principles of Food Sanitation,* 3d ed. Chapman & Hall, New York.

Marriott, N. G. 1982. Foodservice sanitation. In *Professional Development Program for Foodservice Supervisors,* p. VII-1. Academy for Staff Development, Department of Corrections, Commonwealth of Virginia, Richmond.

National Restaurant Association Educational Foundation. 1992. *Applied Foodservice Sanitation,* 4th ed. Educational Foundation of the National Restaurant Association, Chicago.

STUDY QUESTIONS

1. List three basic food safety procedures for foodservice operations.
2. What should a supervisor do if an employee has a cold, flu, or skin infection?
3. List three ways to make sure buffet or cafeteria service is safe.
4. What are four things that increase the risk that food will be contaminated during preparation?
5. Why must workers do cleaning and sanitizing in two separate steps, even if they use a combined detergent-sanitizer compound?
6. How should employees dry equipment, utensils, and surfaces after they wash and sanitize them and why?
7. Give two reasons why sanitation training needs to be ongoing.

TO FIND OUT MORE ABOUT SANITARY FOOD HANDLING IN FOODSERVICE

1. Contact equipment manufacturers or suppliers of cleaning compounds for information about how to clean and sanitize specific pieces of equipment, utensils, or surfaces.

2. The following organizations have books, booklets, posters, videos, and training manuals on sanitary food handling in foodservice:

- The Culinary Institute of America, Video Sales Office, 1433 Albany Post Road, Hyde Park, NY 12538-1499, (800) 285-8280.

- The Educational Foundation of the National Restaurant Association, 250 S. Wacker Drive, Suite 1400, Chicago, IL 60606-5834.

- Food and Drug Administration, 200 C Street SW, HFS-555, Washington, DC 20204, (202) 401-3532.

- Food Marketing Institute, 800 Connecticut Avenue, NW, Suite 500, Washington, DC 20006-2701, (202) 429-8298.

- Colorado State University, Department of Food Science and Human Nutrition, Fort Collins, CO 80523, (303) 491-7334.

- Cornell Cooperative Extension, Department of Food Science, 11 Stocking Hall, Ithaca, NY 14853, (607) 255-7922.

- North Carolina Cooperative Extension Service, Box 7624, Raleigh, NC 27695, (919) 515-2956.

- Ohio State University Extension Service, Department of Human Nutrition and Food Management, Ohio State University, 1787 Neil Avenue, 265 Campbell Hall, Columbus, OH 43210-1295, (614) 292-0827.

- Ohio University, Department of Environmental Health & Safety, Hudson Health Center 212, Athens, OH 45701-2991, (614) 593-0022.

- University of Massachusetts Cooperative Extension, 202 Chenoweth Laboratory, University of Massachusetts Cooperative Extension, Amherst, MA 01003, (413) 545-0552.

- Golden Gate Restaurant Association, 720 Market Street, Suite 200, San Francisco, CA 94103, (415) 781-3925.

- National Environmental Health Association, 720 S. Colorado Boulevard, Suite 970, South Tower, Denver, CO 80222, (303) 756-9090.

- National Live Stock and Meat Board, 444 North Michigan Avenue, Chicago, IL 60611, (800) 368-3138.

- Seattle King County Department of Health, Environmental Health Division, Smith Tower, 506 Second Avenue, Room 201, Seattle, WA 98104-2311, (206) 296-4722.

- Local state and county health departments. Check your telephone directory.

HOW TO AVOID COMMON FOOD SAFETY MISTAKES

In foodservice operations, the most common food safety mistakes are abuse of time and/or temperature, cross-contamination, and poor personal hygiene.

Time/temperature abuse

Foods need to be refrigerated (colder than 2°C, 36°F) or kept hot (hotter than 60°C, 140°F).

- Use digital thermometers with long probes to check the temperature inside the food. Some new thermometers even contain memory chips that keep records of temperatures at different times and on different days. Check oven and refrigerator thermometers.
- Check temperatures of food deliveries to make sure they have been kept cold on the truck. Don't let chilled foods sit out of the refrigerator after they are delivered.
- Cool hot foods in shallow pans using an ice bath or cooling paddle, or use ice as an ingredient instead of water before putting the foods in the refrigerator. Do not cool at room temperature. Putting hot foods in the refrigerator warms up other foods and may not cool the food fast enough.
- Cook foods to temperatures recommended in the FDA Food Code, and reheat foods to at least 74°C (165°F). Do not reheat foods more than once. After cooking or re-heating foods, keep them hotter than 60°C (140°F).
- Prepare foods in small batches. Do not leave large amounts of ingredients sitting out at room temperature while you prepare them.

Cross-contamination

- Buy plenty of color-coded boards, and use each board for one specific type of food—e.g., chicken only, vegetables only, bread only. Wash boards in hot water, and sanitize after each use. If boards go black, they have bacteria growing on them—throw them out!
- Use nonabsorbent, washable mats to anchor cutting boards instead of towels that can absorb juices. Or replace towels between cutting jobs.
- Have specific knives for each type of food; clean and sanitize them between each cutting job. Label the drawers where the knives are kept so that the right knife is used for the right food.
- Wipe down the slicer blade with a clean, hot cloth between jobs, and sanitize. Use an antiseptic block—a block of solidified sanitizer that you slice on the slicer.
- Clean and sanitize the counter between cutting jobs.

- In the refrigerator, store on top shelves cooked foods and foods to be served raw, and uncooked raw foods on bottom shelves, so that contaminated juices do not drip onto ready-to-eat foods.
- On buffets, clean and sanitize or replace tongs, ladles, and spoons every half hour.

Poor Personal Hygiene

- Make sure all employees wash their hands thoroughly and often.
- Make sure employees have convenient handwashing stations so that they are not washing their hands in food preparation sinks (which may contain clean food) or in the washroom (which may have contaminated door handles).
- Train employees to scrub their hands for 30 seconds using antibacterial soap, a nail-brush, and hot water.
- Use plastic gloves properly. Employees must still wash their hands and must change gloves between touching different foods.

Source: Lorenzini, B. 1995. Avoid common food safety mistakes. *Restaurants & Institutions.* 105:122.

Foodservice Control Points

ABOUT THIS CHAPTER

In this chapter you will learn:

1. How to keep food safe at each step in a foodservice facility, from buying and receiving to serving leftovers
2. How to buy, receive, and store food safely
3. How to prepare, serve, cool, and reheat safely
4. How to draw flow diagrams for each recipe used in the facility to help decide on critical control points (CCPs)
5. How to monitor CCPs to make sure each batch of food is safe
6. How the design of the facility and its equipment affect food safety

INTRODUCTION

In a foodservice operation, the flow of food begins when it is bought and delivered (purchasing and receiving) and continues through storing, preparing, cooking, holding, and serving. In a foodservice facility, the sequence may include these steps:

- Buying ingredients
- Receiving ingredients when the supplier delivers them
- Storing ingredients
- Preparing ingredients (thawing, processing)
- Cooking
- Holding or displaying food
- Serving food

- Cooling and storing leftover food
- Reheating and serving food

Each step provides opportunities for food to be contaminated or for bacteria to grow. Food operators need to look at each step to find out what the hazards are (hazard analysis) and how they can make sure they are under control (critical control points, CCPs).

This chapter will discuss how foodservice operators can use control points as food flows through the system to make sure that they always serve safe food to customers.

BUYING INGREDIENTS

It is important to buy foods from reliable suppliers who meet federal and local standards. Managers should be sure that suppliers understand their specifications (detailed descriptions) of the food they want.

Meat and Poultry

All meat and poultry should have a stamp on the carcass, parts, or packaging material to show that the USDA or state department of agriculture has inspected it. Figure 18.1 shows inspection stamps for meats and poultry. Suppliers should be able to show written proof that the government has inspected the meat or poultry.

Shellfish

Shellfish suppliers should be on public-health-service Food and Drug Administration lists of Certified Shellfish Shippers or on lists of state-approved sources. Managers must keep shellstock identification tags on file for 90 days after receiving clams, mussels, and oysters.

RECEIVING AND STORING INGREDIENTS

Employees must check and store food as quickly and efficiently as possible. They must know how to decide whether each food is acceptable and how to take tempera-

FIGURE 18.1. Inspection stamps for meat and poulry: left, inspection stamp for meat; right, inspection stamp for poultry. (Source: U.S. Department of Agriculture.)

tures of refrigerated and frozen foods. Storage areas must be large enough to store foods safely, cleanly, and at the right temperature. Dry storage areas must be clean and dry. Be sure that employees:

- Do not store food in restrooms, furnace rooms, or hallways.
- Do not stack foods on floors.
- Put food on shelves with enough space around them for air to circulate.
- Wrap food in waterproof, nonabsorbing, clean coverings.
- Do not put food directly onto shelves.
- Store above raw foods all cooked foods and foods that will not be cooked any more.
- Follow the "First In, First Out" (FIFO) rule for rotating stock (i.e., use the oldest food first).

Fresh Meat, Poultry, Seafood, and Shellfish

Staff must check the temperature of fresh meat, poultry, seafood, and shellfish when it arrives. Managers should train employees to judge the freshness of the food based on color, texture, smell, and general appearance. Cartons and wrappers must be clean and unbroken.

Frozen foods

All frozen foods, except ice cream, should be −18°C (0°F) or colder when the supplier delivers them. Ice cream may be −14 to −12°C (6°–10°F). An assigned employee should check the temperature by opening one case and putting the sensing area of the thermometer between two packages without puncturing them. A trained employee should look for thawed and refrozen food that has large ice crystals, solid areas of ice, or a lot of ice inside the container. Dried-out food or food that has changed color should be rejected. The packages should be airtight and waterproof.

Canned and Dry Goods

Foodservice facilities should never accept home-canned foods because of the risk of botulism. It is essential to check that cans are not swollen, rusty, or dented, and do not have flaws in the seals or seams. The establishment should return cans if the contents are foamy or smell bad. Employees should never taste the contents to test them, because tiny amounts of bacterial food poison could make them very ill.

Dairy Products

Milk must be pasteurized and should be delivered well before its expiration date. Other milk products, such as cream, dried milk, cottage cheese, soft cheese, and cream cheese, must be made from pasteurized milk. Dairy products should not be sour or moldy.

Employees should check the temperature of milk cartons by opening one and putting in a sanitized thermometer. Employees can check the temperature of bulk plastic containers by folding the soft plastic around the thermometer without puncturing the package. For other products, they should put the thermometer between two packages. Dairy products should be 4°C (40°F) or below.

Eggs and Egg Products

Eggs should be grade AA or A and should have the USDA shield on the carton. Eggs should be refrigerated and less than 2 weeks old. Liquid eggs should be refrigerated or frozen, even though they are pasteurized. Egg shells must be clean and uncracked.

Refrigerated Prepared Foods

Many establishments now buy foods that are partly or completely processed and prepared and can be served to customers with little or no preparation. These foods include refrigerated entrees, prepared salads, fresh pasta, soups, sauces and gravies, and cooked or partly cured meats and poultry dishes. They are sometimes called fresh-prepared foods, fresh-refrigerated foods, or ready-to-eat chilled foods. The foodservice operator has no way to control the safety of these foods (except to store them at the right temperature and use them within the right amount of time). Therefore, the operator should make sure the supplier has a good HACCP program.

If the package has a time-temperature indicator, employees should check it to make sure the food has been stored properly. The right storage temperature is listed on the label of the food.

Foods with Modified Atmosphere Packaging

Modified atmosphere packaging (MAP) is usually used for partially processed foods that are cooked lightly first and then packaged. MAP contains a mix of gases inside the package to lengthen the shelf life of the food. These products may be shelf-stable (heat sterilized: require no refrigeration before they are opened), refrigerated, or frozen. The package may be a pouch, tray, plastic bucket, or glass bottle.

Employees should check the expiration date and/or time-temperature indicator of MAP foods when they are delivered. The gases in the package stop spoilage bacteria from growing, and so the food may not look spoiled but may still have pathogens growing inside. Therefore, it is important to store MAP foods at the right temperature and only until the expiration date.

Sous Vide Foods

Sous vide is a French term that means "under vacuum." Manufacturers put the food in a strong plastic pouch, remove all the oxygen and other air gases, and seal it. Foodservice operators are not allowed to make *sous vide* foods on site, but they can buy them from licensed food processors. *Sous vide* foods may by frozen, refrigerated, or cooked, cooled, and frozen or refrigerated. Uncooked foods are cooked in the sealed pouch by putting it in boiling water.

Frozen *sous vide* foods should be −18°C (0°F) or colder when they arrive. Refrigerated *sous vide* foods should be about 1°C (33–34°F). Higher temperatures shorten the shelf life of the food and may make it unsafe. Again, employees should check temperatures, time-temperature indicators, and expiration dates. Also, they must make sure the package seal is not broken.

Aseptic and Ultrapasteurized Packaged Foods

Aseptic packaged foods are heat-treated to kill pathogens. Foods labeled UHT have been ultrapasteurized (high temperature for a short time) and packaged in a sterile environment. Many UHT foods do not need to be refrigerated, but UHT milk should be 4°C (40°F) or colder.

PREPARING, SERVING, AND REHEATING FOODS

Deciding on CCPs

Most CCPs are in the preparing, serving, and reheating steps. The best way to find the CCPs is to draw a flowchart for each recipe. The flowchart begins with each ingredient and ends with the finished food product or leftovers. Every time the foodservice operation adds a new menu item, managers should draw a flowchart and look for the CCPs. Management should do this when they develop the recipe.

Figure 18.2 shows a recipe for chili. The flowchart in Figure 18.3 shows the flow of food from the time a supplier delivers the ingredients until a customer eats the leftovers. CCPs are shaded. CCPs are any step that needs to be controlled and checked to make sure the food is safe and sanitary. CCPs are different for each food but may include:

- Buying foods from reliable suppliers
- Cooling food within a set amount of time
- Storing food below 4°C (40°F) in the refrigerator or below −18°C (0°F) in the freezer
- Cooking food until the inside reaches a set temperature
- Cooking food for a set amount of time at a set temperature
- Keeping food hotter than 60°C (140°F) or cooler than 4°C (40°F) while it is served
- Not mixing old and new batches of food
- Reheating foods to 74°C (165°F) for a set amount of time

Checking and Monitoring CCPs

It is not enough to decide on the CCPs. Management needs to make sure that all food production meets CCP standards. Managers should set up a record-keeping system. They should design forms that list the CCPs for each recipe, have spaces to record temperatures and times, and have places to log when a CCP is carried out. Employees should complete the forms for every batch of food they prepare.

INGREDIENTS	AMOUNTS
Ground beef	4.5 kg (10 lb)
Onions	2.25g (8 oz)
Tomatoes, canned, diced	2.5L (2 1/2 qt)
Tomato purée	2L (2 qt)
Chili powder	56g (2 oz)
Salt	56g (2 oz)
Pepper, black	1/2 t
Beans, kidney or red, canned	4.25kg (9 lb 8 oz)
Water	1L (1 qt)

PREPARING

1. Chop onions.

2. Drain and rinse beans.

COOKING

3. *Brown beef in stock pot until it reaches 68°C (155°F) for at least 15 seconds.* Drain and add onions.

4. Add tomatoes, tomato purée, chili powder, salt, pepper, beans, and water to stockpot.

5. Heat, stir, bring to boil, then cover and simmer 1 1/2 to 2 hours.

SERVING AND HOLDING

6. Serve immediately.

7. *Hold chili at 60°C (140°F) or higher.* Do not mix new product with old.

COOLING

8. Cool leftovers in shallow pans so that the layer of chili is no more than 2 inches deep. *Cool to 4°C (40°F) or lower within 6 hours.* Stir often.

9. Cover and store at 4°C (40°F) or lower in refrigerator.

REHEATING

10. *Reheat chili to at least 74°C (165°F) for at least 15 seconds within 2 hours. Reheat one time only. Discard any unused food after serving it the second time.*

Note: Italics indicate critical control points (CCPs)

FIGURE 18.2. Chili recipe.

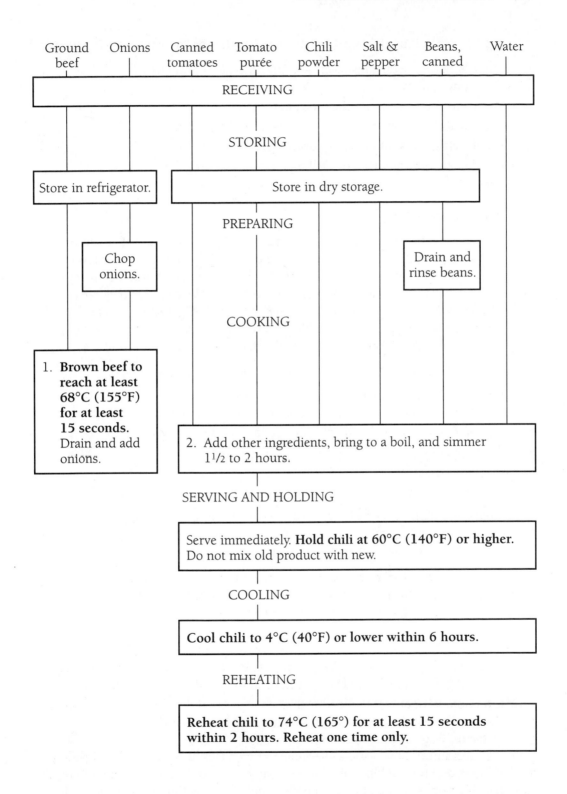

FIGURE 18.3. Chili flowchart. Boldface statements are critical control points (CCPs).

Employees need the right equipment to make sure that they meet CCPs, such as:

- Accurate thermometers; managers should calibrate thermometers at least once a day
- Clocks and timers in convenient locations
- Proper facilities for washing hands, utensils, and equipment
- Enough refrigerator and freezer storage space
- Ice and shallow containers for chilling foods

Standard Procedures

A foodservice operation needs more than just CCPs to make sure that the food is safe. Standard procedures for preparing, cooking, cooling, and reheating foods protect foods from contamination and conditions that would allow pathogens to grow. *All foods that could allow pathogens to grow should be prepared so that they remain less than 4 hours total in the temperature danger zone (4–60°C, 40°–140°F) from the time the ingredients are taken out of storage until the food is served to the customer.* Ways to make sure foods do not spend too much time in the danger zone are:

- Keep refrigerated foods in the refrigerator right until they are prepared.
- Prepare foods as soon as possible before serving them.
- Prepare small batches and keep them in the refrigerator between preparation steps.
- Cool cooked foods quickly by placing shallow pans in ice water.
- Reheat foods quickly and keep them hot during service.
- Serve foods on steam tables or ice beds.

Thawing. Thawing may take several hours or days for large frozen items. It is important to thaw foods safely so that the outside is not warm enough to allow microbes to grow while the inside is still frozen. Choose one of these methods to thaw foods:

- Refrigerate at 3°C (38°F) or colder. Store raw foods on the lowest shelves so that they do not drip or splash onto other foods. Large items, such as turkeys or roasts, take a day or more to thaw (allow about 1 day for each 4–5 lb).
- Cook frozen food to the proper internal temperature (see Appendix 3). This works for vegetables, seafood, hamburger patties, and pie shells, but not for large items, such as turkeys. The cooking time is about 50% longer if the food is frozen (i.e., if it would usually take 1 hour to cook, allow 1.5 hours).
- Use a microwave only if the food will be cooked immediately after it is thawed. This method does not work well for large items, such as turkeys.

Cooking. It is important to use a thermometer to check the temperature inside the food at the end of the cooking time for each batch of food. Employees should not decide that the food is cooked based on how it looks or how long it has been cooking. Other important cooking procedures are:

- Do not overload ovens or cooking surfaces.
- Make sure each portion is the same size and thickness and has the same fat content, so that each one cooks at the same speed.
- Make sure the oven or stove gets back to the right temperature between cooking batches of food.
- Do not touch cooked food with bare hands. Use sanitized serving utensils and cutting boards.

Holding and displaying. Holding equipment should keep hot foods hotter than 60°C (140°F) and cold foods cooler than 4°C (40°F). Other important points are:

- Make sure food is 73°C (165°F) or hotter before transferring it to hot-holding equipment, such as a steam table, double boiler, *bain-marie,* or chafing dish. Do not use holding equipment to cook or reheat food, only to keep it hot.
- Never add new product to old food. Replace the old dish of food with a new dish of food, and replace the serving utensil at the same time.
- Measure temperatures every 2 hours.
- Make sure ice around chilled foods drains away from food containers.
- Stir hot foods often and protect them with lids and sneeze guards.

Cooling and storing. Cooling food the wrong way is the most common cause of foodborne illness. Proper cooling takes time and effort. The temperature inside the food should be below 4°C (40°F) within 4 hours after cooking or the end of hot holding. FDA's 1995 Food Code states that food must cool to 21°C (70°F) within 2 hours, and then to 5°C (41°F) within 4 more hours (total = 6 hours). Cooling needs good planning.

- Use a quick-chill unit, if available, for large food items divided into smaller quantities (such as turkey and roasts), or hot, thick items (such a stews and casseroles).
- Cut large items into pieces. Cut the pieces small enough to cool fast enough. You may need to test how quickly different-size pieces of food cool.
- Use an ice bath or cold-jacketed kettles. Stir or shake foods often while they cool.
- Put the food in shallow pans on the top shelves of a refrigerator. Prechill pans, and use pans made of stainless steel or other materials that transfer heat quickly. Pans should not be more than 10 cm (4 inches) deep, and food should not be more than 5 cm (2 to 3 inches) deep. Do not stack pans, because air cannot circulate around stacked pans.
- Measure the temperature of the food to make sure it is cooling quickly enough. If the food has not cooled within 4 hours, reheat it to 74°F (165°C) or higher for at least 15 seconds within 2 hours, and recool, or throw it out.
- Label foods with the date and time they were prepared, cover, and store.

Reheating. It is important to reheat food thoroughly to 74°C (165°F) within 2 hours of taking it out of the refrigerator so that pathogens in the food are killed and do not have time to grow. Employees must:

- Not reheat food more than once.
- Keep leftover and fresh foods separate.
- Only reheat foods that have been cooled and refrigerated for 2 days or less.
- Never reheat food in hot-holding equipment.

FACILITY AND EQUIPMENT DESIGN

It is hard to maintain good sanitation in a badly designed foodservice facility. Cleaning tasks are difficult and frustrating, and staff waste time and energy trying to clean properly. This often makes workers give up on complete and thorough cleaning.

Cleaning the Facility Easily

Employees must be able to clean and inspect equipment and surfaces easily using normal cleaning methods. Soil, pests, and microorganisms should not have places where they can collect and live. People will be more motivated to clean and keep the facility sanitary if it takes less effort.

Designing the Facility

Operators should consider sanitation when they plan the facility. Most managers work in established foodservice facilities, but they can improve the design if the operator remodels, renovates, or buys new equipment.

Most areas have laws that make sure that foodservice units have a sanitary design. Public-health, building, and zoning departments can all make regulations about the construction of foodservice operations. Regulatory agencies often have checklists to help operators make sure that the building design is sanitary.

Floors, walls, and *ceilings* should be easy to clean and maintain and should also look clean. Operators should use materials that do not react with food or cleaning chemicals, last a long time, do not absorb dirt, and are smooth. If the floor, wall, or ceiling absorbs liquid, the liquid can damage the surface, microbes can grow on it, and it will develop an offensive odor. Food preparation and food storage areas should not have carpets, rugs, or other porous coverings.

The way the floor is built is also important. Coving at the joint between the floor and the wall makes cleaning easier and makes it less likely that bits of food will be left to attract insects and rodents. Operators should seal concrete and terrazzo floors so that they do not absorb liquid or create cement dust.

For walls, ceramic coverings are popular and work well in most areas. Grouting should be smooth and waterproof, and should not have cracks or holes that could collect soil. Stainless steel is expensive but is a good finish because it keeps out water and most soil and lasts a long time. Plaster painted with nontoxic paint or cinder block walls are fine in dry areas, as long as the builder seals them with soil-resistant and glossy paints, epoxy, acrylic enamel, or another good seal. Foodservice operators should not use lead-based or other toxic paints because flakes or chips may get into the food. Ceilings should have smooth surfaces that do not absorb water and are easy to clean. Smooth-sealed plaster, plastic panels, or plastic-coated panels are all good choices.

When the foodservice manager buys new equipment, he or she can make sure that it meets good sanitation standards. Equipment should:

- Have as few parts as possible to do the job
- Be easy to take apart
- Have smooth surfaces with no pits, crevices, ledges, bolts, and rivet heads
- Have rounded edges and corners inside with smooth surfaces
- Be coated with materials that do not crack or chip
- Be made of materials that are not toxic, do not absorb liquids or fats, and do not color or flavor the food

Installing and Laying Out Equipment

Operators should arrange equipment so that food is not easily contaminated and all areas are easy to reach and clean. For example, the dirty-dish table should not be put next to the vegetable preparation sink. Waste and food preparation areas should be as far apart as possible, and food preparation equipment should not be put under an open stairway.

Mobile equipment is best whenever possible, because with such equipment, staff can easily clean the walls and floors around it. Operators should seal equipment that is not mobile to the wall or to other equipment. If they cannot seal the equipment, they should put it 0.5 m (1.5 ft) away from the wall or other equipment so that staff can get around it to clean. Operators should mount equipment that is not mobile about 0.25 m (9 in) off the floor or seal it to a masonry base.

If operators use a seal, it should not be toxic. Operators should not use seals to cover gaps from bad construction, because after a while the cracks will open up and collect soil, insects, and rodents.

Handwashing Facilities

Hands are the most common source of microbes. Therefore, management should provide places for employees to wash their hands wherever hands are likely to get contaminated: in food preparation areas, lockerrooms or dressing rooms, and next to toilets. Employees may not want to walk very far to wash their hands, so handwashing sinks need to be close to where they work. Handwashing sinks should have a washbowl, hot and cold water with foot-operated faucets, liquid or powdered soap, and individual towels or air dryers.

Dressing and Lockerrooms

Employees should have dressing rooms or lockerrooms. Street clothes can be contaminated with microbes, so the operator should provide uniforms for employees to wear while they work. The facility should have a wall or other barrier between dressing rooms and food preparation, storage, and serving areas. Handwashing sinks should be next to the dressing rooms and toilet rooms. Handwashing sinks should not have mirrors above them, so that workers are not tempted to touch their faces or hair after they wash their hands. Staff should scrub the washing facilities and toilet rooms and empty trashcans at least once a day.

Waste Disposal

Garbage and trash disposal need to be done properly, because waste products can attract pests that contaminate food, equipment, and utensils. Trashcans should not leak, should keep out pests, should be easy to clean, and should last a long time. Plastic bags and wet-strength paper bags are good liners for trashcans. Cans must have tight-fitting lids.

Staff should remove trash from food preparation areas often so that it does not start to smell bad or attract pests. Staff should only put waste materials in trashcans. Waste storage areas should be easy to clean and free of pests. If the facility has to keep trash for a long time, the storage area should be indoors and refrigerated. Large dumpsters and compactors outside the building should be on or above a smooth surface that does not absorb liquids like concrete or machine-laid asphalt.

Trash areas should have a supply of hot and cold water and a drain. It is important that water does not contaminate the food preparation or storage areas when staff clean the trash area.

Operators can reduce the amount of trash using a pulper or mechanical compactor. Pulpers grind waste material into pieces that are small enough to flush away with water. The waste is dried, and the processed solid wastes can be trucked away. Compactors squash trash so that it only takes up about 20% of the original amount of space. Compactors are useful in facilities that do not have much space to store trash.

Operators may be able to incinerate (burn) some trash if it is allowed in their area and the incinerator meets federal and local clean-air standards. Most waste from foodservice establishments has too much moisture to burn well.

SUMMARY

- In a foodservice operation, the flow of food has several steps, including buying, receiving, storing, preparing, cooking, holding, serving, cooling, and reheating.
- Operators should buy food from reliable suppliers. Employees should check that the food is fresh, packages are clean and not damaged, and chilled or frozen foods are cool enough when they are delivered.
- The facility must have enough dry storage, refrigerator, and freezer space to stack foods off the ground and with enough space for air to circulate.
- Staff should use the FIFO principle (first in, first out) and should use older food first.
- The flow of food for each recipe can be drawn as a flowchart. This helps identify CCPs.

- Facilities need to monitor CCPs and use good standard procedures to keep food safe.
- The design of the facility and the equipment need to support safe food preparation.

BIBLIOGRAPHY

Bryan, F. L. Hazard analysis critical control point (HACCP) systems for retail food and restaurant operations. *J. Food Protection* 53(11):978–983.

The Educational Foundation of the National Restaurant Association. 1992. *Applied Foodservice Sanitation,* 4th ed. Kendall/Hunt, Dubuque, Iowa.

The Educational Foundation of the National Restaurant Association. 1993. *HACCP Reference Book.* The Educational Foundation of the National Restaurant Association: Chicago.

The Educational Foundation of the National Restaurant Association. 1995. *Serving Safe Food: A Practical Approach to Food Safety.* Kendall/Hunt: Dubuque, Iowa.

U.S. Public Health Service. 1978. *Food Service Sanitation Manual.* U.S. Government Printing Office, Washington, DC.

STUDY QUESTIONS

1. List six of the steps in the flow of food through a foodservice facility. Which step most often causes foodborne illness if it is not done properly?
2. What temperatures should refrigerated and frozen foods be kept at?
3. How quickly must food be cooled to refrigerator temperatures?
4. Foods must be reheated to at least _____ (temperature) for at least _____ (time).
5. What are three safe ways to thaw food?
6. What are three important things about handwashing sinks?

TO FIND OUT MORE ABOUT FOODSERVICE CONTROL POINTS

1. Draw a flowchart for a recipe. Use a foodservice recipe from your workplace or from a textbook. Highlight CCPs for the recipe. Does the recipe need to be prepared differently to make sure the food is safe?
2. Contact the Educational Foundation of the National Restaurant Association, 250 South Wacker Drive, Suite 1400, Chicago, IL 60606-5834, (1-800) 765-2122, for more information on HACCP in foodservice.
3. Use a foodservice-type food thermometer (thermocouple) to practice checking the internal temperatures of food at home.

CHAPTER 19

Management and Sanitation

ABOUT THIS CHAPTER

In this chapter you will learn:

1. Why it is so important for managers to understand good hygiene and support the sanitation program
2. Who managers should involve when developing a sanitation program
3. How managers can help motivate employees and improve the image of sanitation work
4. How sanitation managers can cooperate with the media, outside agencies, and consumers to help the image of the company and promote good relationships
5. What managers need to do to train employees about sanitation
6. Sources of training and education materials
7. How total quality management (TQM) can help the sanitation program

INTRODUCTION

Many unskilled workers choose to work in the food-processing and foodservice industries because they do not need any formal training or education. The food industry often employs high school and college students. Many food industry employees are young and have several outside interests, so employee turnover is high.

Most managers in the food industry agree that sanitation employees do not stay on the job long because they have little training and education and therefore have low salaries. It is difficult for managers to hire and train employees to carry out the sanitation program, in other words to do the cleaning. It is also hard for managers to give sanitation work a professional and exciting image so that employees can be proud and enthusiastic about keeping the operation clean and hygienic. Good managers are the key to a good sanitation program.

MANAGEMENT'S ROLE

If management supports the sanitation program, it is likely to work well. If management does not support the sanitation program, it will fail. The rest of this chapter shows how management is important in designing a good sanitation program and making sure employees carry it out.

Managers' Attitude

Many managers in the food industry do not understand how important it is for a food operation to have an organized sanitation program. Unfortunately, if managers do not think sanitation is important, their employees will not take sanitation seriously. Management may not support sanitation programs because they are expensive and do not quickly increase sales and profits. If top management does not understand sanitation, lower and middle management cannot make it a top priority.

Some management teams are enthusiastic about the sanitation program. They know it can help promote the product, increase sales, and make the food product keep longer. Other managers have improved the image of their organization with good sanitation and quality assurance. Many firms realize that a good sanitation program will save money.

Managers' Knowledge of Sanitation

If management does not understand sanitation and does not support the sanitation program, the facility will not have good hygiene. Managers need to support and promote sanitation and provide an example, because it affects how the operation markets their products. Sanitation programs also affect the company's relationship with inspectors. The U.S. Food and Drug Administration (FDA) and local health department can stop the company from preserving, producing, packaging, storing, or selling any food if their facility or process is unsanitary. Managers need to know that a good sanitation program makes cleaning less expensive because it is more efficient.

Developing the Program

Management should design their sanitation program to fit the operation. In a meat processing operation, many of the conveyors, mixers, and other equipment and containers are open. So the facility needs more hand-held hoses and wands than foam and high-pressure units. A milk plant can use more automatic cleaning equipment such as cleaning-in-place (CIP) systems.

When managers plan the sanitation program, they should get input from the following staff:

1. Employees—especially production supervisors and line workers
2. Quality assurance (QA) personnel—information about areas that need better hygiene and new technology that can improve sanitation

3. Plant engineers—information about equipment maintenance and layouts
4. Purchasing department—how to reduce the cost of equipment and supplies

Managers should give a trained sanitation manager the authority and support he or she needs to make the program work. The sanitation manager should report directly to the plant manager.

The sanitation program should include a Hazard Analysis Critical Control Point (HACCP) program (see Chap. 10). In the future, regulatory agencies may require use of HACCP or a similar program. FDA's Good Manufacturing Practices (GMPs) can help set up a sanitary and hygienic operation because they aim to prevent contamination (adulteration) of foods.

Carrying Out the Program

With good management, everyone involved with sanitation works as a team and shares problems, solutions, and knowledge. For a sanitation program to work, staff need to carry out each step and record the results. Managers can also use an outside sanitation audit to check their sanitation program. Trained auditors have lots of experience and fresh ideas. The sanitation manager or general manager should also carry out an audit regularly. To conform to the HACCP philosophy, sanitation managers should keep a detailed list of problems and what they do to correct them.

Managing a Sanitation Program

Management means "getting things done through people." Sanitation management should get three things done:

1. Delegating sanitation tasks or telling employees what to do
2. Training employees by showing them how to do it
3. Supervising employees to make sure that they do it properly

Managers should inspect the operation regularly to make sure that employees are doing each job properly. Even when managers train employees well, they still need supervision.

Sanitation workers must be able to understand and carry out the sanitation program, and managers must administrate the program properly. Managers cannot afford to make poor decisions about whom to hire or about the sanitation program, because these mistakes could make customers ill. Top management need to understand and support good sanitation and consumer protection.

Management and Supervision

Sanitation programs need good supervision to work properly. Management must audit the sanitation program to make sure that employees are following the rules. Supervisors should always be alert to employees' use of unsafe practices. An ongoing training program reinforces supervision and keeps sanitary habits in the front of employees' minds.

Supervisors need an organized routine to keep track of a food production facility. Managers should supervise food handlers using the same health standards that are used to screen new employees. Many local health codes state that the operator must tell the health authorities at once if they know or suspect that an employee has or is carrying an infectious disease.

The manager's job is easier if employees are motivated to do a good job. Good training can motivate workers. Employees also have better morale and more motivation when supervisors treat them professionally.

Job Enrichment

Many employees, including managers and supervisors, think of sanitation work as a second-rate job. It is important for managers to make sanitation workers aware of the importance of their job. Good management can make sanitation more glamorous and exciting. An effective job enrichment program can make the work more interesting and rewarding for employees. This program can help workers feel that they have a place in the operation. It can demand more of employees, give them more responsibility, and include self-inspection.

Self-Supervision and Self-Inspection

Supervisors may find it hard to set a good example for other employees, but a supervisor who does not follow the rules will not be able to get employees to follow them either.

Trained managers who are familiar with the operation should inspect the establishment often. The owner/operator, managers, supervisors, or sanitation consultants may carry out these self-inspections. Inspectors can use a checklist to make the inspection more valuable and help compare results between inspections.

Public Relations

Food sanitarians should have good public-relations skills and use them in the sanitation program. They must use every reasonable method to help employees understand the program and motivate them to follow every step.

The mass media can help sanitarians promote hygiene in the community in general. If managers have a relationship with news reporters, they can share information and learn to understand each other. Managers can stress improvements, achievements, new programs, new staff, promotions, and other developments. Newspapers, magazines, and television can help the public, the food industry, and food sanitation personnel understand the program better.

Sanitarians should know how to motivate employees. For example, it is easier to motivate groups than individuals. Food sanitarians can do more than inspect the facilities. Other productive and rewarding activities include, for example, talking to a school class or civic group, writing news announcements, helping on radio or TV programs, and making educational materials. It is easier to promote and explain food sanitation when community leaders or civic groups understand and support the idea.

Food sanitarians' recommendations often add to the operating costs of the operation. The sanitation manager needs public-relations skills to sell the need and benefits to other managers.

Cooperation with Other Agencies

Joint advisory committees of regulators, industry managers, and sometimes consumers can evaluate new methods and products. These committees may provide counseling on policies and help form and keep good relationships between regulators, industry, and consumers.

HIRING AND TRAINING EMPLOYEES

Hiring Employees

Managers should be careful when selecting employees who will handle food. They must not have or carry an infectious disease and should have excellent personal hygiene. Good candidates will look neat and clean at the interview.

Training Employees

Good training of employees has been stressed many times in this book. Basic sanitation is a vital part of employee training because no food should be prepared in a food establishment unless everything is clean. Sanitation employees should:

- Be serious, dedicated, and professional
- Understand the company's policy on sanitation
- Know their role in sanitation in the organization

Management, employees, regulatory agencies, and consumers all play a role in food sanitation. In a good sanitation program, sanitation staff work closely with quality assurance (QA) and research and development staff to make sure that standards are met. Staff need to develop the idea of working together, rather than have one department checking up on or policing another. Managers should train all staff to work as a team to keep the facility hygienic, rather than being antagonistic.

Management must hire a well-qualified sanitarian. The sanitarian should be educated about the food-processing operation, cleaning and sanitizing compounds, and food microbiology. The sanitarian should also have experience and/or training in cleaning different types of surfaces. He or she needs to know how to choose the right cleaning equipment, cleaning compounds, and sanitizers for the surface depending on its design, hardness, porosity (whether it soaks things up), vulnerability to oxidation (whether it is damaged by air or oxygen), and susceptibility to corrosion (whether it is eaten away by acid or other chemicals).

A good management team should make sure that the sanitarian knows the following about chemicals used in food sanitation:

- Their safety
- How well they work for specific jobs
- When and how to use detergent auxiliaries and sanitizers
- What type of equipment to use

A sanitarian who understands cleaning equipment, cleaning compounds, and sanitizers can help employees clean efficiently and quickly; waste less time, energy, and cleaning materials; prevent injuries; use less water; and produce less sewage.

When managers give information to sanitation workers, it must be easy to read and understand. Employees should use a clear instruction manual for cleaning all areas and equipment. It should tell employees exactly which cleaning compounds, sanitizers, and equipment to use for each job. The instruction manual should have a sanitation plan and information about pest control, hygienic methods, and maintenance to prevent problems. Use of good hygiene shows in how the facility looks and the quality of the food products, and affects the image of the company.

Customers want and should have wholesome food products. Responsible managers understand that employees need training. These managers have thorough training programs for all employees in their food-processing or foodservice operation. Sanitarians should attend training courses and get help from regulatory agencies to help them train their employees to meet sanitation standards. Management needs to provide training materials, a classroom or area for training, and time for training. The training program is more successful when managers provide good leadership and show enthusiasm for the program.

Other Resources for Sanitation Training and Education

Trade associations and regulatory agencies have information and programs to help managers educate and train employees. Examples are the U.S. Food and Drug Administration, Food Safety and Inspection Service, American Meat Institute, National Food Processors Institute, and National Restaurant Association Educational Foundation. Many of these organizations have instruction manuals for trainers and course books for employees, as well as slides and videos to add interest to the program. Some hold workshops and training sessions around the country. Professional associations such as the International Association of Milk, Food and Environmental Sanitarians help improve the professional status of the sanitarian and educate the food industry about the need for good sanitation programs. Professional associations and universities offer courses on sanitation. A list of names, addresses, and telephone numbers of several of these organizations is shown in Appendix 1.

TOTAL QUALITY MANAGEMENT

Total quality management (TQM) is a management philosophy that allows employees to work with management, have a voice in the operation, and feel some ownership of the firm. TQM is more than a buzz word. This new concept helps management and employees work together for better productivity, lower costs, and more consistent and better products.

TQM and Sanitation

The TQM philosophy applies to management of sanitation. To keep conditions sanitary, each manager and employee needs to take responsibility for keeping the operation hygienic.

In the past, sanitation managers have policed and checked up on employees, instead of having employees be responsible for their own work. TQM involves all employees in decisions and accountability. TQM means offering the customer a consistent and good product through training and hard work by all employees.

In future, food producers are likely to put more emphasis on TQM in their sanitation programs. TQM can help sanitation operations in the same way it has helped manufacturing and service operations. TQM can be a valuable tool for sanitarians.

TQM usually includes a sanitation program for the whole facility, good manufacturing practices (GMPs) for the whole operation, and HACCP program that includes each specific food item.

SUMMARY

- It is very important for managers in the food industry to hire good employees and teach them about good sanitation.
- The sanitation program can only work if management supports it.
- A good sanitation program includes ongoing training and education for employees.
- Education programs can use sanitation training manuals and short courses given by trade associations, professional organizations, or regulatory agencies.
- Sanitation managers decide who will carry out each part of the sanitation program and train and supervise employees.
- Self-supervision and self-inspection are two tools that help keep the operation sanitary.

BIBLIOGRAPHY

Gould, W. A. 1992. *Total Quality Management for the Food Industries*. CTI Publications, Baltimore.

Graham, D. 1992. Five keys to a complete sanitation system. *Prepared Foods* 101(5):50.

Guthrie, R. K. 1988. *Food Sanitation*, 3d ed. Van Nostrand Reinhold, New York.

Holland, G. C. 1980. Education is the key to solving sanitation problems. *J. Food Prot.* 43:401.

Marriott, N. G. 1994. *Principles of Food Sanitation,* 3d ed. Chapman & Hall, New York.

STUDY QUESTIONS

1. Which employees and departments should be involved in developing a sanitation program?
2. What are three things sanitarians should do to manage the sanitation program?
3. What are three ways managers can motivate sanitation workers?
4. Why do sanitarians need to know about cleaning equipment and chemicals?
5. Why do sanitarians need good public-relations skills?
6. Define total quality management (TQM).

TO FIND OUT MORE ABOUT MANAGEMENT AND SANITATION

1. Contact the following organizations for information about sanitation management and employee training:

   International Association of Milk, Food, and
   Environmental Sanitarians (IAMFES)
   502 E. Lincoln Way
   Ames, IA 50010-6666
   (515) 232-6699

   National Restaurant Association Educational Foundation
   250 S. Wacker Drive, Suite 1400
   Chicago, IL 60606-5834
   (800) 765-2122

   USDA Meat and Poultry Hotline
   14th Street and Independence Avenue, SW
   Room 1165-South
   Washington, DC 20250
   (800) 535-4555

   United States Food and Drug Administration (FDA)
   5600 Fishers Lane
   Rockville, MD 20857
   (301) 443-1544

2. Contact a foodservice operator or food processor in your area and ask the sanitation manager about his or her job.

SELF-INSPECTION AND SUPERVISION ARE *So* IMPORTANT

Even small catering and foodservice operations need good hygiene and sanitation practices. In small operations, the owner may need to be the sanitation manager. Regulatory agencies cannot inspect food operations very often, especially small operations, so self-inspection is very important.

In May 1989, 140 guests who attended a catered wedding reception in Napa, California, became ill with *Bacillus cereus* gastroenteritis. The investigation showed that the Cornish game hens had carried the infection.

Spores of *B. cereus* bacteria are found almost everywhere, but large numbers of bacteria have to grow and produce toxin to cause illness. Therefore, *B. cereus* is a rare foodborne illness in the United States. In the Napa outbreak, the bacteria had several opportunities to grow and produce toxin while the food was prepared and transported. Some of the unsafe practices were:

1. The Cornish game hens were not completely thawed before cooking, so the cooking time was not long enough to completely cook the meat.

2. The caterer used the same brush to baste the birds before and after they were cooked. This allowed cross-contamination from the raw to the cooked birds.

3. The outdoor temperature on the day of the event was 90°F, and the caterer transported the hens in an unrefrigerated van for 4 1/2 hours while he made a delivery somewhere else in the county.

4. The caterer used a licensed restaurant kitchen, but the facilities were not adequate for the event.

The investigators recommend that government agencies inspect food safety in catering operations more often. In the meantime, small food operations must make sure they know about food safety and must develop and stick to a rigid sanitation program.

Source: Slaten, D. D., Oropeza, R. I., and Werner, S. B. An outbreak of *Bacillus cereus* food poisoning—are caterers supervised sufficiently? *Public Health Reports* 107(4):477-480.

APPENDIX 1

Where to Get More Information About Food Safety and Sanitation: Agencies, Associations, and Consumer Organizations

The American Dietetic Association (ADA)
National Center for Nutrition and Dietetics (NCND)
216 W. Jackson Boulevard, Suite 800
Chicago, IL 60606-6995
(312) 899-0040 or (800) 877-1600

American Public Health Association (APHA)
1015 Fifteenth Street, NW
Washington, DC 20005
(202) 789-5600

Council for Agricultural Science and Technology (CAST)
137 Lynn Avenue
Ames, IA 50010 7120
(515) 292-2125

The Culinary Institute of America
1433 Albany Post Road
Hyde Park, NY 12538-1499
(800) 285-8280

Egg Nutrition Center
2301 M Street, NW
Washington, DC 20037
(800) 833-EGGS

Drinking Water Hotline
U.S. Environmental Protection Agency
(800) 426-4791

Food Marketing Institute (FMI)
800 Connecticut Avenue NW, Suite 500
Washington, DC 20006-2701
(202) 429-8298

Food and Nutrition Information Center
National Agricultural Library
10301 Baltimore Boulevard, Room 304
Beltsville, MD 20705
(301) 344-3719

Foodservice & Packaging Institute, Inc.
1901 North Moore Street, Suite 1111
Arlington, VA 22209
(703) 527-7505

Hospitality Institute of Technology and Management
830 Transfer Road, Suite 35
St Paul, MN 55114
(612) 646-7077

Institute of Food Technologists (IFT)
221 N. LaSalle Street
Chicago, IL 60601
(312) 782-8424

International Association of Milk, Food,
and Environmental Sanitarians (IAMFES)
502 E. Lincoln Way
Ames, IA 50010-6666
(515) 232-6699

International Food Information Council
1100 Connecticut Avenue NW, Suite 430
Washington, DC 20036
(202) 296-6540

National Automatic Merchandising Association
20 Wacker Drive, Suite 3500
Chicago, IL 60606-3102
(312) 346-0370

National Environmental Health Association
720 S. Colorado Boulevard, Suite 970
South Tower
Denver, CO 80222
(303) 756-9090

National Cattlemen's Beef Association
444 North Michigan Avenue
Chicago, IL 60611
(800) 368-3138

National Seafood Educators
P.O. Box 60006
Richmond Beach, WA 98160
(206) 546-6410

National Restaurant Association Educational Foundation
250 S. Wacker Drive, Suite 1400
Chicago, IL 60606-5834
(800) 765-2122

United States Environmental Protection Agency (EPA)
401 M Street, SW
Washington, DC 20460
(202) 382-2090

United States National Marine Fisheries Service (NMFS)
1335 East-West Highway
Silver Spring, MD 20910
(301) 427-2239

United States Department of Agriculture (USDA)
14th Street and Independence Avenue, SW
Washington, DC 20250
(202) 447-2791

Cooperative Extension Service (CES). Offices are located in most counties.
CES links USDA with state universities. Look in the telephone directory.

Food Safety Inspection Service (FSIS)
USDA Meat and Poultry Hotline
14th Street and Independence Avenue, SW
Room 1165-South
Washington, DC 20250
(800) 535-4555

United States Food and Drug Administration (FDA)
5600 Fishers Lane
Rockville, MD 20857
(301) 443-1544

United States Public Health Service (USPHS)
200 Independence Avenue, SW
Washington, DC 20201
(301) 443-4100

Centers for Disease Control (CDC)
1600 Clifton Road, NE
Atlanta, GA 30333
(414) 639-3286

State and County Health Departments. Offices are located in most counties.
State and County health departments are linked with the CDC, FDA, and other federal agencies. Look in your telephone directory.

A P P E N D I X 2

About Pathogenic Microorganisms

Name of Microorganism (and how to say it)	Sources	Incubation Time (usual time from eating the food to becoming ill)	Symptoms and Signs	Control Points
Anisakis spp. (Ann-is-ah-kiss)	Saltwater fish (e.g., salmon, striped bass, Pacific snapper)	Several days	sore throat, diarrhea, abdominal pain	1. Buy food from reliable sources. 2. Cook thoroughly. 3. Salt heavily. 4. Freeze food at −29°C (−20°F) for 24 hours.
Bacillus cereus (Bah-sill-us seer-ee-us)	Soil, dust, grains, vegetables, cereal products, pudding, custards, sauces, soups, meatloaf, meat products, boiled or fried rice	1–5 hours	Nausea, abdominal pain, vomiting, diarrhea	1. Keep foods hotter than 60°C (140°F) or colder than 4°C (40°F). 2. Chill leftover hot foods quickly to colder than 4°C (40°F). 3. Reheat all leftovers to at least 74°C (165°F) before serving them. 4. Serve and eat foods right after cooking them.
Campylobacter jejuni (Camp-ill-oh-back-ter jeh-june-knee)	Intestines of infected cattle, pigs, chickens, turkeys, and other animals; raw or under-cooked or processed foods of animal origin (milk, poultry, clams, hamburger), unchlorinated water	1–7 days or longer	Diarrhea, abdominal pain, fever, nausea, headache, urinary tract infection, arthritis	1. Cook foods thoroughly. 2. Handle foods hygienically. 3. Dry or freeze food products. 4. Add acids (e.g., vinegar in pickling).

Appendix 2. (*Continued*)

Name of Microorganism (and how to say it)	Sources	Incubation Time (usual time from eating the food to becoming ill)	Symptoms and Signs	Control Points
Clostridium botulinum (Claws-trid-ee-um botch-you-line-um)	Soil, contaminated water, dust, fruits, vegetables, animal feed and manure, honey, sewage, under processed or heated low-acid canned foods, under-processed fermented foods, and smoked fish	12–36 hours	High fever, dizziness, dry mouth, difficulty breathing, paralysis, loss of reflexes	1. Destroy the toxin using a correct combination of time and temperature. 2. Add acids (e.g., vinegar in pickling). 3. Store foods in the refrigerator. 4. Add salts during curing. 5. Destroy all bulging cans and the food inside. 6. Refuse to serve home-canned foods.
Clostridium perfringens (Claws-trid-ee-um per-frin-jens)	Soil, dust, animal manure, human feces, cooked meat and poultry, meat pies, gravies, stews, vegetables that grow in soil (carrots, potatoes, etc.), food cooked and cooled slowly in large quantities at room temperatures	8–12 hours	Sharp abdominal cramps, diarrhea, dehydration	1. Thoroughly clean, cook, and chill food products. 2. Reheat all leftovers to at least 74°C (165°F) before serving them. 3. Keep foods hotter than 60°C (140°F) or colder than 4°C (40°F). 4. Insist on good personal hygiene.
Diphyllobothrium latum (Die-file-oh-bo-three-um late-um)	Freshwater fish (e.g., salmon)	3–6 weeks	Hard to detect (sometimes anemia)	1. Buy food from reliable sources. 2. Cook thoroughly.

Appendix 2. (*Continued*)

Name of Microorganism (and how to say it)	Sources	Incubation Time (usual time from eating the food to becoming ill)	Symptoms and Signs	Control Points
Escherichia coli (Es-cher-ee-chee-ah coal-eye)	Feces of infected people, air, sewage-contaminated water, cheese, shellfish, watercress, ground beef	About 11 hours	Abdominal pain, diarrhea, fever, chills, headache, blood in the feces, nausea, dehydration	1. Heat and chill food products quickly. 2. Insist on good personal hygiene. 3. Control flies. 4. Prepare all food products using sanitary methods.
Listeria monocytogenes (Lis-teer-ee-ah mon-oh-site-oh-jean-ees)	Widespread in nature, contaminated feces, coleslaw, domestic and imported cheeses, chickens, dry sausages (e.g. salami), contaminated meat and meat products	4 days–3 weeks	Mild and flulike headache, vomiting; more severe and can cause death in pregnant women and those with weak immune systems	1. Pasteurize or heat-process food products. 2. Avoid recontamination after heating. 3. Refrigerate or freeze dairy products. 4. Properly clean and sanitize equipment.
Norwalk virus (Nor-walk)	Fish and shellfish harvested from contaminated waters, infected people	24–48 hours	Fever, headache, abdominal pain, diarrhea, vomiting	1. Buy food from reliable sources. 2. Cook shellfish by steaming for at least 4 minutes.

Appendix 2. *(Continued)*

Name of Microorganism (and how to say it)	Sources	Incubation Time (usual time from eating the food to becoming ill)	Symptoms and Signs	Control Points
Norwalk virus (Nor-walk)	Fish and shellfish harvested from contaminated waters, infected people	24–48 hours	Fever, headache, abdominal pain, diarrhea, vomiting	1. Buy food from reliable sources. 2. Cook shellfish by steaming for at least 4 minutes. 3. Insist on good personal hygiene.
Salmonella spp. (Sall-mon-ell-ah species)	Intestines of people and animals; turkeys, chickens, pigs, cattle, dogs, cats, frogs, turtles, and birds; meat products; egg and poultry products; coconut; yeast; chocolate candy; smoked fish; raw salads; fish; shellfish	12–48 hours	Abdominal pain, diarrhea, fever, chills, vomiting, dehydration, headache	1. Cook food products thoroughly. 2. Chill all hot foods quickly. 3. Prevent cross-contamination. 4. Insist on good personal hygiene.
Shigella spp. (Shig-ell-ah species)	Feces of infected people; direct contact with people who carry the disease; contaminated water; uncooked food that is diced, cut, chopped, and mixed; moist and mixed foods (tuna, shrimp, turkey, macaroni, and potato salads); milk; beans; apple cider; contaminated fruits and vegetables	Less than 4 days	Abdominal pain, diarrhea, fever, chills, headache, blood in feces, nausea, dehydration	1. Chill or heat foods quickly. 2. Insist on good personal hygiene. 3. Control flies. 4. Prepare all food products using sanitary methods.

Appendix 2. *(Continued)*

Name of Microorganism (and how to say it)	Sources	Incubation Time (usual time from eating the food to becoming ill)	Symptoms and Signs	Control Points
Staphylococcus aureus (Staff-low-cock-us or-ee-us)	People's noses, throats, hands, and skin; infected wounds and burns; pimples, acne, hair, and feces; cooked ham; poultry and poultry dressing; meat products; gravies and sauces; cream-filled pastries; milk; cheese; hollandaise sauce; bread pudding; fish, potato, ham poultry, and egg salads; high-protein leftover foods	2–4 hours	Vomiting, abdominal cramps, diarrhea, nausea, dehydration, sweating, weakness	1. Do not allow staff to handle food when they are ill. 2. Insist on good personal hygiene. 3. Handle food products with great care. 4. Thoroughly cook and reheat foods. 5. Chill foods quickly and keep them refrigerated.
Trichinella spiralis (Trick-in-ell-ah spur-el-is)	Infected pigs, flesh of bear and walrus	About 9 days	Invades muscles and makes them sore and swollen, weakness	1. Heat pork to at least 66°C (150°F) inside the meat. 2. Store pork at −15°C (5°F) or lower for at least 20 days. (Freezing destroys the parasite.)
Vibrio parahaemolyticus Vib-ree-oh para-heemo-lit-ick-us)	Seawater, raw seafood, saltwater fish, shellfish, fish products, salty foods, cucumbers	10–20 hours	Abdominal cramps, diarrhea, nausea, vomiting, mild fever, chills, headache	1. Cook and chill food products properly. 2. Separate raw and cooked foods. 3. Do not rinse food products with seawater.

Appendix 2. (*Continued*)

Name of Microorganism (and how to say it)	Sources	Incubation Time (usual time from eating the food to becoming ill)	Symptoms and Signs	Control Points
Infectious hepatitis (In-feck-shus hep-a-tie-tis)	Blood, urine, and feces of people and animals who carry the virus; water; rodents; insects; shellfish; milk; potato salad; cold cuts; frozen straw-berries; orange juice; whipped cream cakes; glazed doughnuts; sandwiches	About 30 days	Fever, nausea, abdominal pain, tired feeling, jaundice, liver infection	1. Buy all food products from reliable sources. 2. Insist on good personal hygiene. 3. Cook food thoroughly.
Yersinia enterocolitica (Your-sin-ee-ah enter-oh-coal-it-ah-kah)	Contaminated raw pork and beef, drinking water, ice cream, raw and pasteurized milk, tofu (soy bean curd)	3–7 days	Digestive upset and sharp abdominal pain in children; serious abdominal problems, diarrhea, fever, and arthritis in adults; skin and eye infections in children and adults	1. Pasteurize or heat-process food products. 2. Avoid recontamination of heated foods. 3. Insist on good personal hygiene. 4. Clean and sanitize equipment properly. 5. Buy food from reliable sources.

Safe Cooking Temperatures for Meat & Poultry

Meat and poultry cooked to these temperatures *inside* the meat are generally safe to eat.

Item	Centigrade, °C	Fahrenheit, °F
Fresh Beef		
Rare	60*	140*
Medium	71	160
Ground beef	77	170
Fresh Veal	65	150
Fresh Lamb	65	150
Fresh Pork		
Cooked in microwave	77	170
Cooked using other methods	60	140
Poultry		
Chicken, turkey	65	150
Stuffing (inside or outside the bird)	74	165
Cured Pork		
Ham, raw (cook before eating)	71	160
Ham, fully cooked (heat before serving)	60	140
Shoulder (cook before eating)	71	160
Game		
Deer	71–77	160–170
Rabbit, duck, goose	65	150

*Rare beef is popular, but cooking it to only 140°F means that some food-poisoning organisms may live.

A P P E N D I X   4

Cold Storage Times for Meat and Poultry

This chart shows how long meats and poultry can be stored in the refrigerator or freezer. They may be safe longer, depending on how fresh the meat was when it went into storage, its packaging, and whether it was contaminated. But these times are safe estimates. If food is stored too long in the refrigerator, it may cause foodborne illness. If food is stored too long in the freezer, it will have a bad taste or texture.

Item	Refrigerator, days at 4°C, 40°F	Freezer, months at −18°C, 0°F
Fresh Meats		
Roasts (beef)	3–5	6–12
Roasts (lamb)	3–5	6–9
Roasts (pork, veal)	3–5	4–6
Steaks (beef)	3–5	6–12
Chops (lamb)	3–5	6–9
Chops (pork)	3–5	4–6
Hamburger, ground and stew meats	1–2	3–4
Variety meats (tongue, brain, kidneys, liver, and heart)	1–2	3–4
Sausage (pork)	1–2	1–2
Cooked Meats		
Cooked meat and meat dishes	3–4	2–3
Gravy and meat broth	1–2	2–3
Processed Meats (Frozen, cured meat loses quality rapidly and should be used as soon as possible.)		
Bacon	7	1
Frankfurters	7*	1–2
Ham, fully cooked (whole)	7	1–2
Ham, fully cooked (half)	3–5	1–2

(Continued)

Item	Refrigerator, days at 4°C, 40°F	Freezer, months at −18°C, 0°F
Processed Meats (continued)		
Ham, fully cooked (slices)	3–4	1–2
Luncheon meats	3–5*	1–2**
Sausage (smoked)	7	1–2
Sausage (dry, semidry)	14–21	1–2
Fresh Poultry		
Chicken and turkey (whole)	1–2	12
Chicken pieces	1–2	9
Turkey pieces	1–2	9
Duck and goose (whole)	1–2	6
Giblets	1–2	3–4
Cooked Poultry		
Covered with broth or gravy	1–2	6
Pieces not in broth or gravy	3–4	4
Cooked poultry dishes	3–4	4–6
Fried chicken	3–4	4
Game		
Deer	3–5	8–12
Rabbit	1–2	12
Duck and goose (whole, wild)	1–2	6

*Once a vacuum-sealed package is opened. Unopened vacuum-sealed packages can be stored in the refrigerator for 2 weeks or until the date on the label.

**Freezing not usually recommended.

Glossary

Acid: a substance with a pH between 0 (mostly acid) and 7 (neutral).

Adulterated: food that has been prepared or packed in an unsanitary way, contains dirt, could be harmful to health, or may not be correctly labeled.

Aerobic microbes: microbes that grow only with oxygen.

Agar: growth medium for microbes used during testing.

Alkali: a substance with a pH between 7 (neutral) and 14 (mostly alkaline).

Anaerobic microbes: microbes that grow only without oxygen.

Antimicrobial agent: something that destroys or removes microbes.

Aseptic: free from pathogenic microorganisms.

Aseptic packaging/aseptic canning: food and containers are commercially sterilized separately, cooled, and containers filled and sealed in aseptic conditions.

Bacteriostats: substances or conditions that slow growth of bacteria.

Bactericides: substances or conditions that kill bacteria.

Biological control: control of pests or microorganisms using biological methods, e.g., hormones to interfere with reproduction or natural predators.

Biofilm: microcolonies of bacteria that stick to surfaces using a matrix and trap debris and other microbes.

By-product: something other than the main food product that is produced during food processing; may include waste products or wastewater that can be recycled or reused in some way.

Carrier: someone who harbors and releases a pathogen but does not have symptoms of the disease.

CCP: Critical Control Point: a step where control (time, temperature, or protection) can remove a hazard, prevent a hazard, or lessen the risk of a hazard.

Centralized kitchen: one large kitchen facility that prepares foods to be served at several locations.

Chambers test (also called the sanitizer efficiency test): the sanitizer should kill 99.999% of 75 million to 125 million *Escherichia coli* and *Staphylococcus aureus* bacteria within 30 seconds at 20°C (68°F).

Chelating agent (sequestering agent or sequestrant): an inorganic compound that is blended with cleaning compounds to prevent deposits of salts from hard water (i.e., scale).

Chemical poisoning: food poisoning caused by chemicals that are in the food by accident.

CIP: Cleaning in place: the equipment design includes tanks for cleaning and rinse solutions and spray heads and drains inside the equipment to clean it without taking it apart; works well for fluid processing, e.g., dairy and beverages.

Clean: free of visible soil.

Cleaning: removal of soil from a surface.

Colony-forming unit (CFU): units used to count the number of bacteria in a given area or volume; a colony is a collection or group of bacteria derived from one or a few microorganisms.

Commercial sterility: not completely sterile, but enough bacteria and spores are destroyed for the food to be safe.

Communicable disease: infectious disease.

Concentration: the strength of a solution, the amount of compound dissolved in a given volume of liquid.

COP: Cleaning out of place: workers take apart equipment or move it from the processing area to the cleaning area to clean it; works well for small parts and utensils.

Corrosive: eats away, wears away, or weakens a surface.

Critical zone (see danger zone): the range of temperatures at which microbes grow fastest, 2–60°C (36–140°F).

Cross-contamination: carrying soil, microbes, or other particles from a raw-food product to a finished-food product on equipment, utensils, clothing, air, or other vehicles.

Danger zone (see critical zone): the range of temperatures at which microbes grow fastest, 2–60°C (36–140°F).

Delegation: giving someone responsibility for a job.

Disinfection: killing harmful microorganisms.

Dispersion: breaking up material into fine particles that are suspended in solution.

Effluent: waste material discharged from a processing plant.

Enzyme: a biological catalyst that speeds up chemical reactions.

Facultative microbes: microbes that grow either with or without oxygen.

FIFO: first in, first out: principle for rotating stock so that older product is used before the newer inventory.

Food infections: illnesses caused by eating infectious microorganisms, such as bacteria, viruses, or parasites.

Food poisoning: an illness caused by eating food that contains toxins made by microbes or chemical poisons.

Food spoilage: changes in the color, flavor, or texture of foods.

Foodborne infection: illness caused by growth of a foodborne microorganism in a person's body.

Foodborne intoxication: illness caused by toxins produced by microorganisms in food before the food is eaten.

Foodservice: providing food for a large number of people in a commercial operation (restaurant, fast-food operation, or cafeteria) or institution (school, college, military operation, hospital, residential care, prison, etc.).

Fumigant: a vapor or gas used to destroy pests.

Fumigate: using a gas or vapor to destroy pests. A bomb that releases gas is placed in the room, and the room is closed up for several hours to allow the gas to work. The area is well ventilated before people go in again.

Gastroenteritis: inflamed stomach and intestines leading to vomiting, diarrhea, and cramping.

Germicide: chemical that kills microorganisms.

Homogeneous: the same composition or consistency throughout.

Host: the food, animal, or plant that supports growth of a microorganism.

Hygiene: practices used to maintain good health.

Inert gases: gases that do not react with other substances.

Infestation: when pests live somewhere and cause disease and destruction.

Inhibitors: substances or conditions that prevent growth of microbes.

Injunction: a court order that requires a company to correct a problem or stop production.

Inorganic: chemicals or substances that are not made by plants, animals, or other living organisms, e.g., minerals, salts.

Insecticide: chemical poison that kills insects.

Integrated pest management (IPM): effective, environmentally sound, and economical control of pests using a combination of methods (inspection, housekeeping, physical and mechanical methods, biological methods, and chemicals).

Job enrichment: a program to make work more exciting and rewarding for employees.

Low-moisture foods: foods that contain very little moisture, such as bakery goods, nuts and seeds, candy, cereal, grains, and snack foods.

Mechanized: automatic equipment.

Meningitis: inflamed and/or infected membranes around the brain.

Mesophiles: microbes that grow best in moderate temperatures (20-45°C, 68-113°F).

Microbe, microorganism: organisms too small to be seen with the naked eye, such as bacteria, viruses, and some parasites and yeasts.

Microbiology: study of microscopic forms of life, i.e., bacteria, viruses, etc.

Mode of transmission: how the pathogen is carried from the source to the food.

Modified atmosphere packaging (MAP): foods with inert gases inside the sealed package; the gases slow down food spoilage, but pathogens can still grow.

Motivation: giving employees a reason to do their jobs thoroughly.

Mycotoxin: toxin produced by a mold.

Nutrients: substances needed by organisms to grow.

Organic: chemicals or compounds with a carbon skeleton made by plant, animal, or other living beings.

Outbreak: an incident in which two or more people have a similar foodborne illness.

Parasite: an organism that lives off another organism, e.g., parasitic worms that live in the intestines of humans or animals.

Pasteurization: partial sterilization of a food (usually a liquid) at a temperature and for a period of exposure that destroys pathogens without major chemical changes in the food.

Pathogen: a microorganism that can cause disease.

Perishable: a food that spoils or decays quickly or easily, e.g., fruits, vegetables, fresh meat, or milk.

Personal hygiene: maintaining a clean body and clean clothing.

Pest: an animal that causes destruction or disease (e.g., rodents, insects, and birds).

Pesticide: chemical poison that kills pests.

pH: a measurement of acidity and alkalinity, ranging from 0 to 14: values <7 are acidic, 7 is neutral, and values >7 are alkaline.

Pollution/pollutants: contamination of the environment with waste products causing harm to soil, water, plants, and animals.

Porosity: being porous, ability to soak up fluids.

Processing: making a food product from raw ingredients; may include butchering, cleaning, chopping, cooking, mixing, etc.

Productivity: rate of production or amount of food produced per hour.

Psychrotrophs: microbes that grow best in cold temperatures, below 20°C (68°F).

Public health: the protection and improvement of community health by organized community effort, including preventive medicine and sanitary and social science.

Quality assurance: planned and systematic inspection, monitoring, evaluation, and record keeping to maintain an acceptable product.

Quarantine: isolation of a person, animal, or plant to prevent spread of disease.

Quats: quaternary ammonium compounds used as sanitizers.

Ready-to-eat: food that will not be heated or processed again before it is eaten.

Regulations: detailed rules, written after an act has been passed, that specify how and when the law will be carried out.

Relative humidity: availability of water in the atmosphere.

Rodenticide: chemical poison that kills rodents (rats and mice).

Sample: a part of something that represents the whole (e.g., a small amount of milk can be tested to find out about the contents of the whole container).

Sanitary: free of disease-causing microorganisms and other harmful substances.

Sanitation: creating and maintaining hygienic and healthful conditions.

Sanitation audit: formal examination of an operation's sanitation program.

Sanitizer: process or agent (heat, radiation, or chemicals) that kills or removes microbes.

Sanitizing: killing or removing all microbes.

Seafood: edible fish and shellfish.

Shelf life: how long a food product is good to eat before it contains dangerous levels of microbes or the color or flavor goes bad.

Shellfish: water animal with a shell—molluscs (e.g., mussels, clams, snails) or crustaceans (e.g., lobster, shrimp, crab).

Soil: material in the wrong place, often made up of dirt, dust, and specs or scraps of food.

Source: where a pathogen comes from.

Sous vide **foods:** foods in a strong plastic package from which air is removed (i.e., under vacuum).

Specifications: details of a product or process, e.g., amount, ingredients, temperatures, etc.

Spoilage: decay; off-color, flavor, or odor; or other damage that makes food unacceptable to eat.

Spore: an inactive, resistant, resting, enclosed cell that can produce another growing organism when the conditions are right.

Stability: resistance to chemical, physical, or microbial breakdown.

Statistically valid sample: a large-enough size and number of samples to be sure that it represents the entire batch.

Sterilization: removal or killing of all living organisms, especially microorganisms.

Superheating: heating equipment or food long enough at a high enough temperature to destroy pests, insects, microbes, molds, fungi, and/or spores.

Supervision: making sure employees do each step of their job.

Surfactant: a complex molecule that is blended with a cleaning compound to reduce the energy of the bonds around the soil and allow closer contact between the soil and the cleaning compound.

Suspension: the process of loosening, lifting, and holding soil particles in solution.

Terminal sterilization: the heating step in the traditional canning process. Food is poured into containers, sealed, and then heated to destroy most bacteria.

Thermophiles: microbes that grow best in hot temperatures, above 45°C (113°F).

Total quality management (TQM): a management philosophy that helps managers and employees work together for better productivity, lower costs, and more consistent and safer products.

Toxin: a poisonous chemical.

Ultrapasteurized foods: foods pasteurized at a high temperature for a short time to kill pathogens; the foods are packaged aseptically and may be labeled UHT (ultrahigh temperature).

Vegetative cells: active, living cells.

Wastewater: water contaminated with waste from food processing and cleaning.

Water activity (A_w): availability of water in a food or beverage.

Water hardness: the amount of inorganic salts (such as calcium chloride, magnesium chloride, sulfates, and bicarbonates) in water.

Water softening: removes or inactivates the calcium and magnesium ions in water.

Wetting (penetration): caused by a surfactant that allows the cleaning compound to wet or penetrate the soil deposit and loosen it from the surface.

Wholesome: promoting health or well-being; safe.

Answers to Study Questions

CHAPTER 1

1. Sanitation means creating and maintaining hygienic and healthful conditions.
2. Substantive, which means they have the power of law and must be carried out.
3. The Food and Drug Administration (FDA). Meat and poultry products are not monitored by this agency.
4. Less expensive; better communication between regulators and firms.
5. Conflicting interests of other departments; monitor sanitation across the organization as a whole.
6. Meets regulatory requirements; protects the reputation of the brand and products; ensures that products are safe, of high quality, and free from contamination.

CHAPTER 2

1. Molds, yeasts, bacteria, and viruses. (See chapter for examples.)
2. Yeasts look slimy and creamy white; molds are fuzzy and cottonlike.
3. Lag phase—adapting to the environment.

   Logarithmic growth phase—cells multiply rapidly.

   Stationary growth phase—various factors limit growth.

   Accelerated death phase—lack of nutrients and buildup of waste products cause rapid death of microbial cells.

   Reduced death phase—the number of microbes decreases, so the death rate slows.
4. Molds.
5. Temperature, oxygen, relative humidity, water activity, pH, nutrients, and inhibitors.
6. They are harder to remove than normal dirt and microbes. They provide a place for microorganisms that cause spoilage and foodborne illness to grow. Chunks of biofilms can break off into food during processing.
7. Physical spoilage is obvious because of changes in the color, texture, smell, or flavor of the food. It may be aerobic (with oxygen) on the food surface or anaerobic (without oxygen) inside the food or in vacuum-packed or canned foods. Chemical spoilage is when enzymes in the food and from microorganisms break down protein, fats, carbohydrates, and other large molecules in the food into smaller and simpler compounds.
8. *Salmonella* spp., *Staphylococcus* spp., *Clostridium perfringens, Clostridium botulinum, Campylobacter* spp., *Listeria monocytogenes, Yersinia enterocolitica,* and *Escherichia coli* O157:H7.
9. Cereal grains, flour, bread, cornmeal, popcorn, peanuts, and peanut butter. Avoid insect damage and keep dry.
10. Destroying microbes means killing them so they cannot grow. Inhibiting microbes means injuring them or creating conditions in which they do not grow well; the microbes can recover and, if conditions become good, they can start to grow again.

CHAPTER 3

1. Infection—microorganism in food grows in the body after it is eaten.

   Intoxication—microorganism grows and produces toxins in the food before it is eaten.
2. Good source of protein, amino acids, B vitamins and minerals.

   Handled extensively.

   Frequently stored for long periods without refrigeration.
3. Food production and foodservice employees.
4. Disposable gloves, frequent handwashing.
5. Air filters, appropriate packaging.
6. Steam table, ice tray, transparent shield over and in front of food.
7. Covered, in clean containers, and away from food storage areas.

CHAPTER 4

1. Bathe daily, scrub fingernails daily, wash hair twice a week, wear clean clothes. Get enough sleep, eat well, and avoid exposure to infectious diseases when possible to cut down on illness.
2. At the beginning of their shift; after using the toilet; after handling garbage or anything dirty; after handling raw meats, poultry, eggs, or dairy products; after handling money; after smoking; and after coughing and sneezing.
3. Everyone carries bacteria that can cause foodborne illness, even when they are healthy. Employees can carry diseases even if they do not have symptoms or have recovered.
4. Skin, hands, fingernails, hair, mouth, nose, waste organs.
5. Jewelry is a safety hazard if it is caught in machinery. Jewelry may be contaminated and may fall into or come in contact with food.
6. Set a good example, educate and train employees, check employees health and skin condition, inspect employees work habits, provide good toilet and handwashing facilities, provide disposable gloves, provide good laundry facilities.

CHAPTER 5

1. Smooth, hard, nonporous.
2. Step 1: Separating soil from the surface or equipment.

   Step 2: Dispersing soil in the cleaning solution.

   Step 3: Preventing dispersed soil from reattaching to the surface.
3. Breaks up soil into smaller particles or droplets that can be dispersed and carried away by the cleaning solution; stops dispersed soil from settling.
4. Food particles protect microbes from sanitizers.
5. Free of minerals (i.e., soft water), free of microorganisms, clear, colorless, noncorrosive.
6. Alkaline cleaners remove organic soils, e.g., fat, oils, grease, protein. Acid cleaners remove inorganic soils, e.g., minerals, encrusted mineral scale.

7. Away from traffic and food supplies; off the floor; in a cool, dry, locked, fire-safe cupboard; acid and alkaline cleaners separated; in clearly labeled, sealed containers.

8. Flush the skin or eyes with plenty of water for 15 to 20 mintues.

CHAPTER 6

1. Make sure they are thoroughly clean.

2. Killing or removing all microbes.

3. Heat, radiation, chemicals. Chemicals are used most often in food-processing facilities.

4. Expensive, uses a lot of energy, inefficient, easy to mistake water vapor for steam, can make soil cake onto surface, need to check that the temperature stays hot enough.

5. Fruits, vegetables, and spices.

6. The temperature of the solution must be less than 55°C (131°F), and the soil must be light.

7. Destroys all types of vegetative bacteria, yeasts, and molds quickly.

    Works well in different environments (i.e., soiled surfaces, hard water, different pHs, soap or detergent residues).

    Cleans well.

    Dissolves in water.

    Stable as purchased (concentrate) and as used (diluted).

    Easy to use.

    Readily available.

    Inexpensive.

    Easy to measure.

    Does not irritate skin, nontoxic.

    No offensive odor.

8. Chlorine: Works on a broad range of microorganisms, cheap, no rinsing required at 200 ppm or less, available as liquid or granules, not affected by hard water.

    Iodophors: Stable at low pH, most effective sanitizer for viruses, can be used at very low concentrations, not affected by hard water, stain unremoved organic and mineral soils, do not irritate skin.

    Quats: Penetrate well, natural wetting agents, Good for *L. monocytogenes* and mold, stable in presence of organic soil, not corrosive or irritating, heat stable.

    Acid: Neutralizes excess alkalinity from cleaning, can combine with final rinse, does not corrode metals, works against yeasts, stable when heated and in presence of organic matter.

CHAPTER 7

1. Labor.

2. Mechanized cleaning systems, reuse water and cleaning solutions, lower temperatures, use the right amount of chemical at the right concentration, use the right chemical for the type of surface or equipment to reduce corrosion, good overall management.

3. Fragments can get stuck and can corrode surfaces or contaminate food.

4. Fog makes it hard to see, condensation can encourage mold growth, steam can burn workers.

5. The force helps clean the equipment or surface and gets the cleaning solution into hard-to-reach areas.

6.

Centralized	Portable
Advantages	
Outlets are available at several places in the facility. Reduces labor costs.	Workers can move the equipment to any area that needs to be cleaned. Equipment is fairly cheap to buy and maintain. Works well against *Listeria monocytogenes*.
Disadvantages	
Equipment is expensive to buy and maintain. System must be custom designed.	Takes more labor to move equipment to the area where it is needed. Less automated than centralized units. Less durable than centralized units.

7. Foam clings to the surface, so the cleaner does not need to be reapplied several times, and the worker can see where the foam is and is unlikely to clean the same area twice.

8. CIP equipment is not taken apart; built-in heads spray cleaning and rinsing solutions into the equipment, and built-in drains remove and/or recirculate the solutions; works well for fluid processing (e.g., dairy and beverages), does not work well for heavy soil; system needs to be custom designed and is expensive to buy and maintain; a microprocessor unit can control the operation. COP equipment is taken apart and cleaned in another area; works well for heavily soiled equipment; less expensive than CIP systems to buy and maintain; works well for small parts and utensils.

CHAPTER 8

1. It attracts insects and rodents, smells unpleasant, becomes a public nuisance, and looks unattractive.

2. Microbes grow on the food and use up the oxygen dissolved in the water. When the amount of dissolved oxygen in the water falls below 5 ppm, the fish die.

3. BOD (biochemical oxygen demand) is the amount of oxygen waste uses as it decomposes.

4. Checking water balance, sampling wastewater, and checking the amount of pollution.

5. BOD (biological oxygen demand), COD (chemical oxygen demand), DO (dissolved oxygen), TOC (total organic carbon), SS (settleable solids), TSS (total suspended solids), TDS (total dissolved solids), and FOG (fats, oils, and grease).

6. Truck to municipal garbage dump, compost, dry and grind for animal feed.

7. Advantages: Grease and solid materials recovered from waste can be sold, may reduce municipal surcharges; fewer complaints from the municipal treatment facility.

    Disadvantages: Pretreatment facilities are expensive and make the processing operation more complex; wastewater treatment can be expensive to maintain, monitor, and document; pretreatment facilities may be subject to property tax.

8. Pretreatment reduces the levels of pollutants in the wastewater to meet municipal regulations. Primary treatment removes particles from the wastewater. Secondary treatment oxidizes dissolved organic matter. Tertiary treatment removes pollutants such as colors, smells, brines, and flavors. Disinfection removes most of the remaining microbes from the water.

CHAPTER 9

1. Cockroaches, houseflies, and fruit flies.

2. Pests carry microorganisms in and on their bodies. They transfer them to food, surfaces, and equipment when they walk on them, bite or chew them, spit saliva onto them, or leave droppings on them.

3. Walk into a dark room and turn on the lights; check for a strong, oily odor; or look for their small, black or brown, spherical feces.

4. Good sanitation, fill cracks, inspect deliveries for adult cockroaches and eggs.

5. Keep flies out using air screens, mesh screens, double doors, and self-closing doors. Remove garbage promptly, and keep it covered and enclosed so that it does not attract flies. Use electric traps.

6. Residual pesticides work for a while after they are applied. Nonresidual pesticides only work for a short time when they are used.

7. Prevent them from entering; make sure there are no places for them to live; remove food sources; kill them by gassing, trapping, poisoning; tracking powder.

8. • Make sure all pesticide containers have clear labels.
    • Check that exterminators have work insurance.
    • Follow instructions when using pesticides.
    • Use the weakest poison that will destroy the pests.
    • Use oil-based and water-based sprays in appropriate places.
    • Wear protective clothing while applying pesticides, and wash hands afterwards.
    • Be careful not to contaminate food, equipment, and utensils with pesticides.
    • Call a doctor or poison control center if accidental poisoning occurs.

9. Use of a combination of several methods to control pests effectively and economically with minimal use of pesticides and minimal effects on the environment. IPM includes inspection, housekeeping, mechanical and physical methods, and chemical and biological methods.

CHAPTER 10

1. To make sure that the firm produces safe and acceptable food while maintaining productivity and efficiency.

2. a. Clear objectives and policies.

   b. Sanitation requirements for processes and products.

   c. An inspection system that includes procedures.

   d. Specified microbial, physical, and chemical values for products.

   e. Procedures and requirements for microbial, physical, and chemical testing.

   f. A personnel structure, including an organizational chart for the QA program.

   g. A QA budget to cover related expenses.

   h. A job description for all QA positions.

   i. An appropriate salary structure to attract and retain qualified QA staff.

   j. Constant supervision of the QA program and regular reports of the results.

3. Inspect facilities and equipment, develop specifications and standards, sample and test products, check staff hygiene, meet government regulations, evaluate cleaning programs, report results, train staff, troubleshoot and correct problems, keep accurate records.

4. Exact, e.g., temperature, pH. Subjective: e.g., flavor, smell.

5. A critical control point is any point in food production where microorganisms should be destroyed or controlled to prevent a hazard. A control point is any point during food production where loss of control may result in an economic or quality defect, or the low probability of a health risk.

6. Because microbial tests often slow, and results may not be available for hours or even days. Microbial tests may be used to make sure the HACCP program is working properly, i.e., verification.

CHAPTER 11

1. *Listeria monocytogenes.* Survives aerobic and anaerobic conditions, grows at refrigeration temperatures, found in animal intestines and some plants; contamination can come from several sources, can survive pasteurization if contamination is heavy.

2. Water supply, waste disposal and drainage facilities, energy supply, proximity to highways, proximity to milk supply, proximity to a large populated area.

3. Minerals, lipids, carbohydrates, proteins, and water. Dust, lubricants, microorganisms, cleaning compounds, and sanitizers.

4. Presoften the water or use additives to the cleaning compound to overcome the hardness.

5. Save labor and energy; only a small amount is needed.

6. Heat products for the minimum amount of time at the minimum temperature possible, cool product heating surfaces before and after emptying processing vats, and keep soil films moist by rinsing away foam and other debris and leaving water in the processing vats until cleaning.

CHAPTER 12

1. *L. monocytogenes* can grow at refrigerator temperatures.
2. • Scrub floors and drains every day. Rinse drains with disinfectant every day.
   • Scrub walls each week.
   • Clean the outside of all equipment, light fixtures, sills and ledges, piping, vents, and other processing and packaging areas that are not part of the daily cleaning program.
   • Clean cooling and heating units and ducts each week.
   • Caulk all cracks in walls, ceilings, and window and door sills.
   • Reduce condensate on ceilings.
   • Scrub and clean raw-material areas as often as the processing and packaging areas.
   • Keep hallways and passageways clean and dry.
   • Minimize traffic in and out of processing and packaging areas. Establish plant traffic patterns to reduce cross-contamination from feet, containers, pallet jacks, pallets, and fork trucks.
   • Change outer clothing and sanitize hands, or change gloves when moving from an area containing raw ingredients to an area containing finished product.
   Change into clean work clothes each day at the plant. Color-code clothing to show which clothes belong in each area.
   • Make sure visitors change into clean clothes provided at the plant.
   • Monitor the surfaces and air to make sure that *Listeria* are under control.
   • Enclose processing and packaging rooms so that they only receive filtered air.
   • Clean and sanitize all equipment and containers before bringing them into processing and packaging areas.
3. • Microorganisms change the color and flavor of meat and poultry products.
   • Longer shelf life for self-service merchandise.
   • Less meat and poultry is wasted because it is discolored or spoiled.
   • Sanitary conditions improve the image and reputation of a firm.
   • Regulatory agencies and consumers insist on good sanitation.
   • Sanitary and tidy surroundings improve morale and productivity of employees.
   • More processing and handling of food require a stricter sanitation program.
   • Sanitation is good business.
4. Pick up debris, prerinse and wet, wash with cleaning agent, rinse, inspect, sanitize, and prevent recontamination.
5. CCP1—skinning: The hide is very contaminated, and there is no good way to remove soil from the live animal before slaughter. Therefore, it is important to skin the animal to cut down on cross-contamination from the hide to the carcass.
   CCP2—prewash: This step should remove most of the contamination from contact with the hide and minimize attachment of microbes to the carcass.
   CCP3—bacterial wash: Bactericidal compounds (e.g., acetic acid) reduce contamination with microbes, including pathogens from the animal's intestines.

CCP4—evisceration: Animals hold large numbers of pathogens in their intestines. The best way to make sure these pathogens do not contaminate the carcass is to teach operators how to remove the intestines and internal organs without bursting them.

CCP5—final wash: This bactericidal rinse reduces the number of microbes on the carcass and minimizes the number of pathogens that are carried into the rest of the processing and packaging process.

CCP6—chill: Rapid chilling controls the growth of pathogens.

CCP7—storage: Strict temperature control and daily cleaning and sanitizing of equipment prevent pathogens from growing. The population of pathogens at this point shows how well the previous CCPs have controlled contamination. Maintaining the product temperature below 7°C (45°F) is a CCP to prevent growth of pathogens during packaging and distribution.

6. Changes in supervisor, training new employees, complicated procedures may be hard to remember.

CHAPTER 13

1. Chilling begins immediately after harvesting. Chilling reduces the temperature of the product to 10°C (18°F) within 4 hours. Chilling continues to approximately 1°C (34°F).

2. a. Cover electrical equipment with polyethylene sheeting.

   b. Remove large debris and put it in trash containers.

   c. Manually or mechanically remove soil from the walls and floors by scraping, brushing, or hosing with mechanized cleaning equipment. Start at the top of equipment and walls, and work down toward the floor drains or exit.

   d. Take the equipment apart.

   e. Prerinse with water at 40°C (104°F), or lower to wet the surfaces and remove large and water-soluble debris.

   f. Apply a cleaning compound that works on organic soil (usually an alkaline cleaner) using portable or centralized high-pressure, low-volume or foam equipment. The cleaning solution should not be hotter than 55°C (131°F).

   g. Leave the cleaning compound for about 15 minutes to work on the soil, and rinse the equipment and area with water at 55 to 60°C (131-140°F).

   h. Inspect the equipment and the facility, and reclean if necessary.

   i. Make sure that the plant is microbially clean by using a sanitizer.

   j. Avoid contaminating the equipment and the area during maintenance and setup.

3. • Good public-image first impressions for inspectors and the public.

   • The condition of the outside of the plant often reflects the plant's hygiene standards inside.

   • Wet areas may contaminate food products through seepage and by providing a place for microorganisms and insects to grow.

   • Very dusty roads, yards, or parking lots can contaminate areas where food is exposed.

   • Untidy refuse, litter, equipment, and uncut weeds or grass around the plant buildings may be a good place for rodents, insects, and other pests to breed.

4. • Use of uncontaminated wastewater from one area of a food-processing operation (e.g., water from the final rinse in a cleaning cycle) in other areas that do not need drinkable water

  • Use of closed water systems in food-processing operations in which all water used in processing is continuously filtered to remove solids

  • Use of dry conveyors to transport solids, instead of water

5. Pathogens in seafood are rare and testing methods are not very accurate or precise. HACCP methods are less expensive. HACCP identifies and controls factors that could cause a hazard, rather than just checking to see if the end product is safe.

CHAPTER 14

1. Heat.
2. Soil, air, pests.
3. Easier to keep clean, fewer employee injuries.
4. So that pests and microbes do not have places to live or enter the building.
5. High-volume plants, juice processors.
6. Traditional: Cans are filled, sealed, and then heated to sterilize. Aseptic: Food and containers are sterilized separately, food is cooled, then containers are filled and sealed in aseptic conditions.
7. So that he or she knows which tests to use.

CHAPTER 15

1. They are mostly sugars and dissolve in water.
2. There is no good way to detoxify a beverage after it is made.
3. a. Rinse to remove large particles and soil that is not stuck on, to wet the area, and to help the cleaning compound work properly.
   b. Apply a cleaning compound (usually as foam) to wet and penetrate the soil.
   c. Rinse to remove the dispersed soil and the cleaning compound.
   d. Sanitize to destroy microorganisms.
   e. Rinse again if using a quaternary ammonium sanitizer before the cleaned area touches any beverage products.
4. Flocculation, filtration (through a sandbed), chlorination, sterile filtration, reverse osmosis, activated carbon, and deionization
5. Advantages: Penetrates well, no chemical residues. Disadvantages: Expensive, can leave condensation, slower than chemicals.
6. Bottling.
7. Acidity, carbon dioxide content, sugar level, alcohol level.

CHAPTER 16

1. Foods that contain very little water, such as bakery goods, nuts and seeds, candy, cereal, grains, and snack foods.

2. Insects, rodents, mold.

3. Dust provides food for rodents, birds, insects, and microbes and can get into finished food products.

4. Control humidity and temperature, have good ventilation, inspect and clean raw ingredients and storage and processing areas regularly, make sure the building is well maintained so that pests cannot get in, keep the inside and outside of the building tidy.

5. Advantages: Kills pests; can be used to treat a piece of infested equipment. Disadvantages: Uses a lot of energy, expensive.

6. Hand brooms, push brooms, dust and wet mops, dustpans, vacuum cleaners (portable or installed), mechanical scrubber or sweeper, compressed air, cylindrical brushes.

CHAPTER 17

1. Wash food, protect food, keep food hot or cold.

2. Give them tasks where they do not touch food, or do not let them work until they are healthy.

3. • Cooling tables to keep foods at 2°C (36°F) or cooler.

    • Warming tables to keep foods at 60°C (140°F) or hotter.

    • Authorized workers to serve food in hygienic conditions; customers not to handle serving utensils.

    • Servers to wear hairnets or hats, and not to touch hair while serving.

    • Transparent shields between customers and service tables and dishes.

4. Multiple preparation steps, temperature changes, large volume, and contaminated basic foods.

5. The sanitizer can be neutralized by the dirt during cleaning, so it needs to be used after cleaning for it to be able to sanitize.

6. Air-dry, because wiping cloths can recontaminate them.

7. Employees need reminders to keep sanitation a top priority; employee turnover is high in the foodservice industry.

CHAPTER 18

1. Buying, receiving, storing, preparing, cooking, holding, serving, cooling, reheating. Cooling causes foodborne illness most often.

2. Refrigerated below 4°C (40°F); frozen below −18°C (0°F).

3. Within 6 hours.

4. 74°C (165°F), 15 seconds.

5. In refrigerator, cooking from frozen, in microwave.

6. Close to work stations, foot-operated faucets, powder or liquid soap, and individual towels or air dryers.

CHAPTER 19

1. Managers, quality assurance (QA), research and development, plant engineers, purchasing department, sanitation employees.

2. Delegate, train, supervise.

3. Training, professional treatment, job enrichment program, total quality management (with employee input); use good hygiene themselves.

4. To help employees clean efficiently and quickly; waste less time, energy, and cleaning materials; prevent injuries; use less water; and produce less sewage.

5. To "sell" sanitation to managers and employees, to help the company have a good public image.

6. TQM helps management and employees work together for better productivity, lower costs, and more consistent and more hygienic products.

INDEX